Contents

Introduction

About Riches of the Earth

Much of the industrial expansion of the nineteenth and early twentieth centuries in the South Pennines was only possible because of what lay under the ground here. The Industrial Revolution was of huge global and national significance and it was through the extraction of local sources of coal, good quality building stone, and other raw materials, that made the South Pennine uplands of key importance.

Pennine coal fuelled mills and steam engines that powered industrial machinery; miners working up here were literally fuelling the fires of industrialisation. Pennine stone was used for the construction of roads, mills, houses, and other structures, and fireclay was extracted to make heat-proof bricks, able to withstand the high temperatures of industrial boilers.

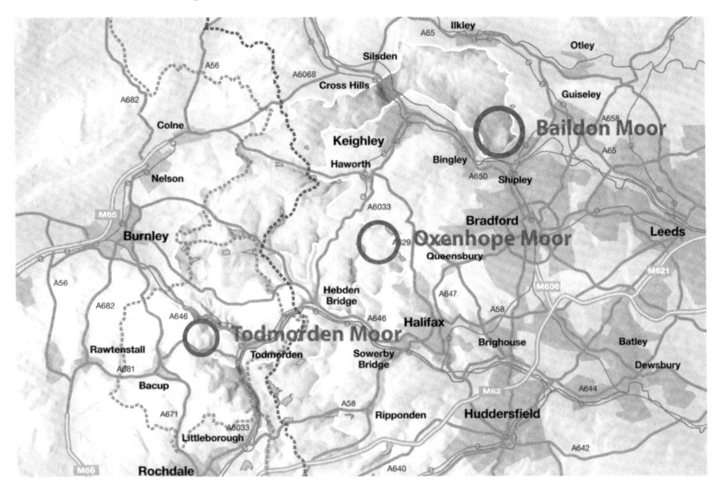

But what is the legacy of the industries that were once present in the South Pennines moorland landscape? How do we know what went on? How can we tell what impact the extraction industries had on the landscape of the South Pennines and of its people?

The aim of the 'Riches of the Earth' project was to investigate three areas - Baildon, Oxenhope and Todmorden Moors - to uncover the secrets of the extraction industries and stories of what went on beneath our feet. Three survey teams recorded what survives and researched available written documents and oral history evidence. By drawing evidence together, we may begin to understand these important industries and the legacy they have left behind.

Baildon Moor

Team members: Derek Barker, Rosemary Cole, Tony Grogan, Mike Lawson, Malcolm Leyland, Eddie Nash, Dave Shepherd, Paul White.

Baildon Moor lies to the north of Baildon, the River Aire and the Leeds and Liverpool Canal. Baildon Moor has a rich cultural history spanning at least 4,000 years. A number of prehistoric cup-and-ring marked stones are sited on the moor, as well as the remains of military training and defences relating to World War II. However, it is the moor's rich geology - stone, coal and clay - that has played a part in shaping the surrounding settlement. Features on Baildon Moor have been surveyed through a programme of fieldwalking and detailed survey of a selection of features. The Riches of the Earth team was drawn from individuals with an interest in the industrial remains on the moor who met together back in September 2010. Further evidence has been drawn together from documentary, photographic and oral history sources.

Oxenhope Moor

Team members: Duncan Boud, Josh Granger, Chris Mace, Angela Speight, William Varley.

Nab Hill, which makes up part of Oxenhope Moor, lies on the watershed of the Rivers Aire and Calder in West Yorkshire and on the boundary of the present Metropolitan areas of Bradford and Calderdale. The road which cuts through the area was once the packhorse route between Haworth and Halifax. The area now consists of a series of long-abandoned ruined farmhouses and scattered disused quarry workings. The quarries exploited flagrock present in the escarpment edge and on the plateau beyond for building stone. Survey work on Nab Hill comprised a detailed metric and photographic survey of one area of quarrying alongside extensive documentary and historical research. A small group of individuals came together to undertake the survey under the supervision of Chris Mace, a local archaeologist who once quarried close to the survey area.

Todmorden Moor

Team members: Ian Booth, Linda Kokoska, Ian Lever, Robin Pennie, Sarah Pennie, Sue Turner.

The rich moor at Todmorden lies to the west of the town. The moor was exploited for coal, clay and sandstone. A small group of volunteers from the Todmorden Moor Restoration Trust undertook the survey and historical study of the mineral extraction evidence on Todmorden Moor. Evidence has been drawn together from aerial photography, fieldwalking, and documentary, photographic and oral history sources.

Circular feature with raised bank surrounding a central depression, indicating the remains of a mining shaft. Todmorden Moor

An artist's impression of what Carboniferous forests would have looked like

A Lesson in Stone

The geology of the South Pennines is dominated by the sandstone and mudstones of the Millstone Grit and Coal Measures, with a small area of Carboniferous Limestone to the north-west. This section focuses on the Carboniferous sandstones, as the three survey areas, Baildon, Oxenhope, and Todmorden, all lie within this area, and it is this rich mineral resource that has been exploited.

Carboniferous times

The rocks which form the landscape of the South Pennines were formed during the late Carboniferous period, between 315 and 305 million years ago. It is hard to imagine that most of the area which is now Britain was a lowland plain which enjoyed a hot, rainy tropical climate, with mountains lying to the north and south.

Sediment brought down from the mountains by rivers was deposited in estuaries or deltas, in an environment similar to the present-day Mississippi or Ganges deltas. Wide shallow river channels flowed between sandbanks surrounded by flat plains, which were occasionally flooded when rainfall was very high.

Surrounding the river channels were low-lying areas with lakes, marshes and lagoons. Muds were deposited in the still waters of lakes and swamps, and were covered up from time to time by sand brought by floodwaters.

Later, in the Carboniferous period, the continent was above sea-level for longer, so forests thrived and coal seams were formed. Rocks deposited during these times, from 310 – 300 million years ago, are called Coal Measures.

Fossils

Because the climate was warm and rainy, forests grew on the surrounding lowlands. Plant fossils are common in sandstones, as branches and tree trunks were carried by floods and were left stranded on river sandbanks as the water receded.

How a goniatite would have looked when it was living in Carboniferous seas

Mudstone beds occasionally include fossils of marine creatures, such as brachiopods and bivalves (shellfish). Goniatites are the coiled ancestors of the ammonites. The sea level must have risen occasionally to flood low-lying land, bringing sea creatures with it. When they died, their shells were fossilised in mud layers. It is often difficult to find fossils because of the vegetation cover, although they can be found in some mine spoil tips on Baildon Hill. On Todmorden Moor, goniatites can be found in carbonate nodules, uncrushed and very well-preserved.

The Pennine anticline

At the end of the Carboniferous period, the continent was uplifted and tilted during a major collision between two tectonic plates. This mountain-building period is called the *Variscan orogeny* and culminated in the uplift of a high mountain range

A calcareous concretion found on Todmorden Moor, with a small goniatite next to it

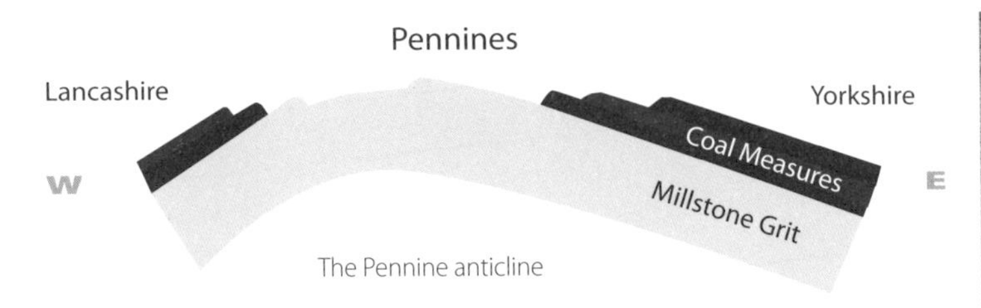

The Pennine anticline

across Europe. Northern England was uplifted into the Pennine anticline, which is an upfold. For millions of years after the end of the Carboniferous period, the present Pennines formed a range of hills trending north-south.

Sandstones, siltstones and mudstones

Sandstones are rocks which are made of sand grains up to 2mm in size. If you look at a fresh piece of sandstone with a hand lens, you will see grey quartz grains. Siltstones are like sandstones, but are made of very fine quartz particles. Quartz is a resistant mineral, so sandstones often form the ridges, scars and high moors of the Pennines. Millstone Grit is the name given to rocks of this age because the sandstones are rough and were suitable for making millstones for grinding grain.

Mudstones, often called shales, are made from mud. The mud is made up of clay particles, complex minerals with a variety of chemical compositions. Mudstones often produce good soils which hold water and provide nutrients for plants. They are rarely visible at the surface as they weather rapidly to form a soil layer which is then colonised by vegetation. Mudstones are sometimes visible in moorland gullies on steep slopes, where heavy rain has exposed the solid rock.

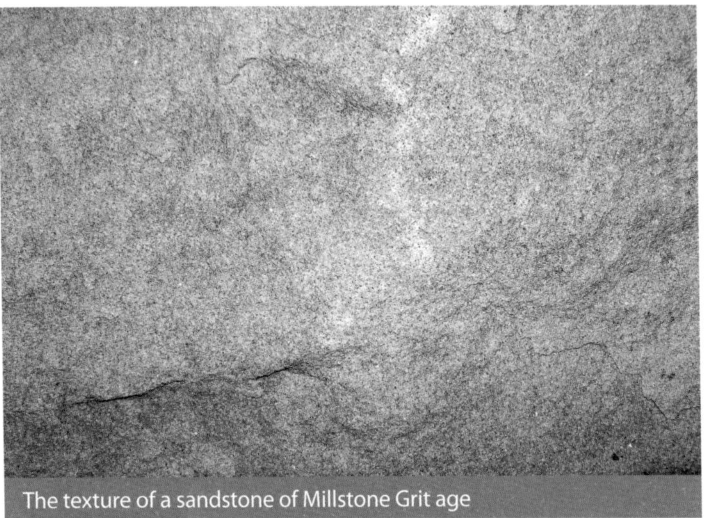
The texture of a sandstone of Millstone Grit age

Laminated black mudstones, interbedded with bands of orange ironstone nodules at Swan Bank, Halifax

How rocks are made from sediments

The sands, silts and muds that form the rocks of the South Pennines were deposited in shallow water by rivers. Subsequently, they were buried by many metres of sediment and compressed so that any water was squeezed out. Water moving through the sediments carried minerals which cemented the sand and mud grains together to make sandstones and mudstones; sedimentary rocks.

Quartz, calcite and iron oxides are the most common cementing minerals for sandstone, filling up the spaces between the sand grains by crystal growth. When iron is present in the cement, the sandstone takes on a red, yellow or brown colour. Quartz cement gives a rock increased resistance to weathering, whereas a large proportion of iron oxides in the cement reduces the strength of the rock. Sometimes, orange holes or voids are visible in quarry faces. This is where the cement is not strong enough to bind the sand grains together, so weathering of the rock takes place more easily.

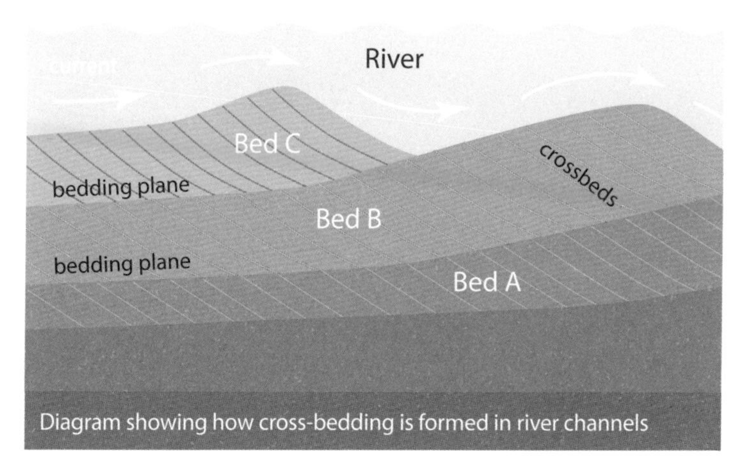

Diagram showing how cross-bedding is formed in river channels

Structures in sandstones and mudstones

Sedimentary rocks are made up of layers or beds. Most sediments are deposited in rivers, seas or lakes so the bedding of the rocks is close to horizontal and is called regular bedding. However, if you look at many quarry or cliff faces in the South Pennines, you can see that many of the lines on the rocks lie at an angle to the regular bedding. This is called cross-bedding and is associated with the formation of sandbanks in large river channels which carried sands and muds down from the mountains during Carboniferous times.

Mud particles are so fine that they settle out only when the water in lakes or swamps is not moving. The thin beds that can be seen in mudstone (shale) gullies on the moors are called laminations.

A thin coal seam (black) underlain by grey fireclay in Dimples Quarry, Haworth. The scale is 8cm long

Coal, fireclay and ganister

During Carboniferous times, the continent which is now Europe was close to the equator and the climate was warm and rainy, rather like the present Amazon basin. Forests grew luxuriantly but plant species were very different.

When branches and leaves died, they fell into stagnant water in swamps and lakes, where they decayed slowly to form a black organic sludge. Sediments were laid down above the organic material as river channels deposited more sand. Water, oxygen and hydrogen were driven out of the plant remains by the weight of wet sand. Carbon from the plants remained to form coal seams.

Underlying the coal seams on Baildon Hill and Todmorden Moor there are layers of yellow/grey clay. This is fireclay, which is a type of seat-earth. Seat-earths represent the soils in which vegetation grew and sometimes contain thin, black rootlets. Fireclays are normally less than 30cm thick, though on Baildon Hill and Todmorden Moor some fireclays are recorded as being up to three metres thick.

Ganister is also a seat-earth and is a hard, fine-grained, quartz-rich sandstone formed from sandy soils found in Carboniferous forests. Quarry workers extracted ganister to be crushed and mixed with fireclay to create manufacturing material for lining steel furnaces. Ganister was also called 'galliard' or 'calliard' by quarrymen in Yorkshire and Lancashire.

Ironstone

If you pick up a stone which is ochre or dark red/brown and feels particularly heavy, it is probably an ironstone nodule. They are found as bands of nodules within mudstones. They often appear to have concentric layers which sometimes peel off easily. The proportion of iron in ironstone nodules was high enough to make them worth exploiting, so iron smelting took place in Medieval times close to Baildon Moor and in many other locations across West Yorkshire.

Ganister is a hard, pale-coloured sandstone. It often contains fossil plant rootlets preserved by black carbon

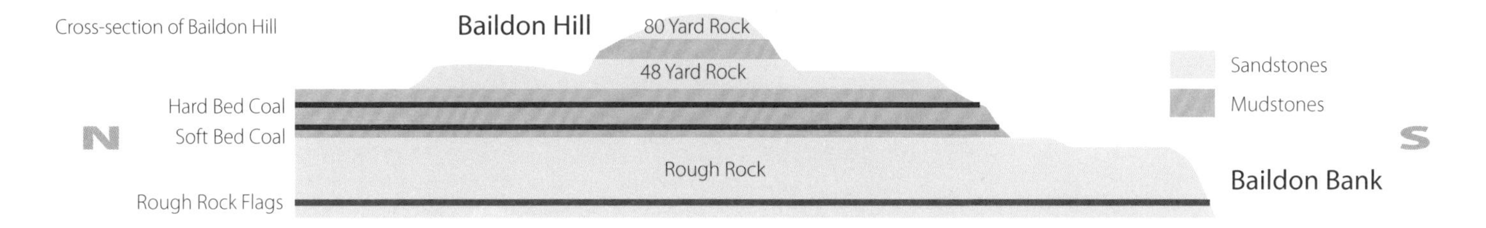

Cross-section of Baildon Hill

Baildon Moor

Baildon Hill is made of horizontal lower Coal Measures beds of sandstones and mudstones. The '80 Yard Rock' forms the summit and the '48 Yard Rock' forms the main platform which runs around the hill, but it is not easy to find exposures of these sandstones. Their names come from the old miners' measurements from the base of each rock to the Hard Bed Coal – the most important coal seam in the area.

The Millstone Grit sandstones, of which the Rough Rock and Rough Rock Flags are the most important in this area, are older than the Coal Measures rocks. They outcrop around the base of Baildon Moor. The Rough Rock is a tough, resistant gritstone, quarried extensively to the east of Baildon Hill, at Eaves Crag and at High and Low Eaves Delves, as well as at Baildon Green and Baildon Bank quarries to the south of Baildon Hill.

The best stone, with well-spaced, regular bedding, is used for construction, particularly in the lower walls of large buildings. Stone with bedding spaced more closely together is used for flagstones or roofing slates. Kerbstones, setts and building stone would have been masoned to the right size by stone-masons working in the quarries. Any rocks which had weaknesses, such as plant fossils, would have been used for field walls. Waste stone was sometimes crushed and used for tracks and paths. However, much waste stone was also left in spoil tips close to the quarries.

Nab Hill, Oxenhope Moor

Nab Hill is a broad, moorland plateau on the watershed between the Worth and Calder valleys, with its highest point at 415m. Nab Hill has always been a conspicuous landmark, more so since the building of a large wind farm on its summit. This gently sloping plateau with steep slopes to the north and west reflects the underlying geology, which consists of layers of sandstones and intervening mudstones. The more resistant sandstones form the top of the plateau, the weaker mudstones are more easily eroded and so form the steeper slopes beneath. The main sandstone is the Rough Rock Flags which have been quarried since the nineteenth century and used for flags, roofing slates and building stone. The rock was extracted using hand tools from the small quarries or delphs; small hand cranes were probably the only form of mechanisation used. The delvers cleared 'a baring', which involved removing the surface overburden to get to the good flag rock beneath. On the top of the moor, shafts were also sunk to get down to the layers of good stone beneath the surface.

In the 1870s there were 15 quarries in operation. Quarrying continued until after World War I. Today the abandoned delphs remain, together with the extensive tips of waste rock gradually being reclaimed by vegetation.

Todmorden Moor

Todmorden Moor is an upland plateau moor at about 350 to 400 metres above sea level, lying between Cliviger Gorge to the north and Dulesgate to the south, with steep slopes into the Calder valley.

The moor is cut by five different sandstone beds, which run roughly north-south and dip down towards the south-west. If you walk up Flower Scar Road, you can see exposures of sandstone in the track. There are good exposures in two disused quarries close to the Bacup Road (A681).

Between each sandstone bed lie mudstones containing a total of seven coal seams, that have been extensively worked below Todmorden Moor. Mudstones are exposed in the small gullies in the moorland, as well as in the spoil tips and along the line of the old tramway near Sandy Road Colliery.

Many of the spoil tips around the mine contain marine fossils, such as goniatites, within carbonate nodules. The coal seams and associated mudstones are an important source of coal balls which are of great importance in the study of Carboniferous plants.

Coal balls were misnamed by miners who found them in the coal seams. However, they are not made of coal, but are nodules made of calcite (calcium carbonate) which contain beautifully preserved plant fossils, in which each individual plant cell has been replaced by calcite.

A coal ball from Todmorden Moor showing pale-coloured plant fossils

Clough Head sandstone and fireclay quarry

Extraction in the South Pennines

Scratching the surface

Some areas of the South Pennines are covered in peat, an organic deposit which has slowly grown in depth over the thousands of years. In other parts of the South Pennines, in what is known as 'Millstone Grit country', the underlying rocks are visible as scarps and outcrops. The rock layers contain 'special ingredients' that have been sought out through need and opportunity for hundreds of years.

That layers poke out of the ground so obviously meant that stone and coal could be easily extracted. Simply digging at the ground's surface was all that was needed to win coal or break out rock. Iron made by Medieval monks using Baildon Moor ironstone, family houses warmed with coal dug from the Lord of the Manor's pits, and houses built from Oxenhope stone were all made possible through material found on or near the surface.

Extraction techniques evolved so that coal and ironstone were tunnelled out in 'drifts' which chased horizontal seams. Later methods saw miners digging shafts straight down into a seam where material was extracted in all directions until it got too dangerous to continue. This method formed what are known as 'bell pits'. When one bell pit was finished it was filled with waste material and another shaft dug nearby down to the same seam. This technique resulted in the remains of concentrations of bell pits in one area. Techniques developed further to include the extraction of material to form whole underground rooms. Material was excavated horizontally creating galleries with ceilings supported by pillars of coal that could be dug away as miners retreated to their exit.

A different method of extraction was seen in the north-western area of the South Pennines. Limestone is found here on Worsthorne Moor in glacial till (boulder clay), in the form of cobbles or boulders embedded in clay brought by ice-sheets from the Clitheroe area to the north. The cobbles were extracted from the glacial till by 'hushing' (see page 14), a process in which torrents of water flooded through the deposits, breaking up the clay and separating the larger boulders. The limestone boulders were cleaned in water, burnt in lime kilns and transported as lime. Spreading agricultural lime has always been valuable for soil improvement as it reduces the acidity of soils formed from sandstones, but limestone is also used as a flux in many industrial processes.

Digging deep

There were many reasons why deposits nearer the surface were worked over a long period of time. Even when bigger and better quality seams were known to exist, deeper mines presented a huge range of lethal dangers, such as continual flooding and underground gases. Adding to these practical problems was the cost implication of excavating deeper mines. For a long time it was simply not worth the investment for what you might hope to expect in return. As a result coals and stones stayed as local products for centuries.

Demand for building stone and fuel coal from further afield made deep mining a worthwhile investment during this country's Industrial Revolution. In the eighteenth and nineteenth centuries, goods were transported to markets ever-further away, in ever-greater quantities. New towns nearby were growing, and South Pennine stone flags might have been sold in London or the American colonies.

A delph on Nab Hill with quarrying areas in the foreground and ruins of dressing sheds and waste tips beyond

Limestone hushing at Sheddon Clough, Worsthorne. The blue arrows indicate how water travelled across the landscape leaving large cuttings behind

By the nineteenth century almost every settlement in Lancashire and West Yorkshire had been growing rapidly for nearly 100 years because of the textile trade. Now the textile mills were powered by hungry steam engines. The seven coal seams deep under Todmorden Moor came to be worked in earnest to supply the nearby mill engines and millworkers' fireplaces, and Baildon Moor coal was taken to fuel steam engines in textile mills as far as Otley. Coal meant riches for the men and companies able to supply it in quantity. One of Britain's earliest transport mega-projects, the Bridgewater Canal, was designed and built to bring down the cost of getting coal from the Duke of Bridgewater's coalfields in Lancashire to his prospective customers in Manchester. Later on, the railway network would also use steam-powered engines riding on wrought-iron bullhead rails which had developed out of the iron and coal industries. Mines then sprang up close to rail tracks with their own depots.

Moving on

Coal is still used in huge quantities worldwide, stone is still used for building, and iron ore is still used to make metals. The cost of extracting material from the South Pennine moorland areas became greater than from elsewhere because these upland places were relatively remote and inaccessible. Material gained from extraction could be obtained cheaper elsewhere, particularly from well-connected parts of the country that were well-served by one of the many railways criss-crossing English soil.

It was not a simple rise and fall though. Money and effort was still ploughed into the moorland operations by smaller companies with limited numbers of people working above or below ground. Stone was still quarried, coal mined and even clay, used to make drain pipes, began to be taken from Todmorden in the middle of the 1800s. The real action was simply focused elsewhere.

Baildon Moor was not mined as extensively from around the beginning of the twentieth century, and mining was actively discouraged by the Bradford Act of 1899. No miners are recorded in the 1901 census although it is possible that some may have travelled from further afield to work in the very small number that continued to operate as a local concern. Presumably some of the skilled workers had to go elsewhere to find similar work. The Oxenhope quarries were too remote to become fully mechanised. Quarry machines were probably used from the late 1800s but quarrying using manual labour continued into the twentieth century. At Baildon and Todmorden coal continued to be taken for private use until after World War II, but by the 1960s all the remaining collieries and machinery had simply been abandoned.

What remains in the uplands of the industries and of these people's lives? The mining complexes and quarry buildings were demolished and cannibalised, or else they became fossilised as they were abandoned. Surviving maps show us where things were at a given point, and there are a handful of photographs and plans. Some of the plans show the layout of mines as they were abandoned but there are far less abandonment plans than there were mines, and mining companies were not required to produce much detail about them. Nonetheless we have still inherited a selection of tips, bumps, shafts and ruined buildings which all form patterns in the landscape. By looking at what survives, as well as the few surviving documents, it is possible to start to piece together how the moors were exploited, who the characters were that worked there, and what life was like for them.

Extraction Communities

Settling in...

Records show us how fast places around the moorlands grew from the end of the eighteenth century onwards as a result of the extraction industries. Hamlets grew around mines and delphs. New cottages and terraced houses were sometimes built to house workers while some farms were bought up by mining companies, and temporary shacks thrown up around the existing farm buildings.

Censuses and parish records report where people were born and what they did. From patterns in the data conclusions may be drawn about the impact extraction had on communities, and how the operations were organised and run.

Status quo

Mines, quarries and pits grew amongst and out of communities that had farmed the moorlands since Domesday. The right to take material from a landowner's property had probably developed out of the old customary landowner-tenant rights, which allowed tenants to take from land owned by a far-off landlord. In the main, mine and quarry Masters seem to have been well-connected local men, lynchpins of the community who could exercise control over a workforce and arrange for goods and wages to be distributed. Sometimes Masters were also pub landlords, doubling up on their existing roles as employers, distributors and merchants. At first the mining and quarry workers were drawn from the local community. Later on, experienced workers from elsewhere were drafted in, perhaps as things got more competitive or as new concerns were established. The newcomers often brought their families with them, and census records reveal where they came from.

Surviving records demonstrate that family ties and kinship were important in extraction communities. Women and children played very important roles both at home and at work. Women and children certainly worked in the mines, carting the coal from the working face to the shaft where it would rise to the surface drawn by horse or steam-engine power. Children worked as soon as they were considered able. The very youngest would work as 'trappers', pulling the heavy curtains open to let the coal tubs pass; some old miners could recall being afraid of the dark from an early age.

The Children's Employment Commission of 1842 found that children were working in the mines from the age of five, and the commission considered them even worse off than children working in factories. Women miners in the West Riding were found to be working naked to the waist, and men completely so. Parliament was appalled at the commission's findings and, despite some opposition, their legislation sought to regulate child labour and keep women and girls above the surface. Women and children under the age of 10 were eventually forbidden to work in coal mines by the Mines Bill of 1842, but many women cross-dressed in order to continue earning a miner's wage. This was not a great wage, but it was still better than an agricultural wage, or indeed nothing. The appointment of an Inspectors of Mines by Act of Parliament in 1850 was no doubt intended to show that the authorities were serious about addressing these issues.

Remains of Fly hamlet between Nab Hill and Warley Moor reservoir

Women, sometimes but not always widows, also leased rights to mine or quarry from at least as early as the eighteenth century. Widows would also act as stone merchants, grocers and beersellers on the death of their husbands. Their sons would work the delph if the father's lease was inherited by surviving family members.

What a life

The extraction industries have always presented dangers to the lives of the workers and average life expectancy, particularly among miners, has usually been below the average. The work was back-breaking and required much stooping under the ground and dragging heavy loads over the surface. Every day there were long shifts with little rest, and food and drink was sparse. Workers could be run over, crushed by rockfall, or slip down a shaft to their death. For miners there was also the constant danger of being blown up. Explosions were caused by badly laid charges, or by methane, known as 'firedamp' or 'minedamp' by miners. Methane gathered in pockets and was easily set off to explode by a naked flame or by a spark from spontaneously combusting coal spoil heaps. Sir Humphrey Davy's safety lamp was invented in 1815. This hid naked flames behind a mesh which prevented methane combustion. Sadly this invention increased the number of accidents at first, when sections of mines previously closed off for safety reasons were re-opened.

Children were particularly vulnerable to accidents. Workers as young as six could easily be maimed or killed, and they often were. It was not until the mid-1800s that newspapers and government studies began to document accidents and fatalities at work, and it is likely that conditions may have been much worse than those documented.

All change

The eighteenth and nineteenth centuries represented an increasingly chaotic age. The country changed beyond recognition, physically and socially. Traditional age-old rights to take fuel and stone from the landlords' property were lost as wealthy families and then companies bought up extraction rights. Other traditional ways of life were stamped out, too. The mechanisation of textile production in huge mills killed off the hand-weaving that had supported generations of upland farming communities. The opening of the Bradford Canal in 1774 helped the South Pennines economy to expand but it also connected it physically and socially to the wider world, opening up the horizons for families and individuals who would previously have been very strongly tied to the places where they were born.

As things continued to become more commercial and competitive, new blood was eventually brought into the South Pennines to ensure that extraction outfits were run efficiently and at a healthy profit. Documentary records show how the extraction industries continued to grow in importance until recent times, and how reliant communities became on the income it generated. When extraction sites were shut down, some of the old timers continued to take material in small quantities until they were physically unable. The scars in the landscape, the abandoned machinery and the mothballed buildings reflect some of the changes these industries have wrought in the moorland and on the surrounding communities.

Quarryman at the entrance of a nineteenth century adit driven underground to exploit flagrock. Photograph taken in the mid-1980s

Survey 1: Discoveries on Baildon Moor

Introduction

Baildon Moor is a three square kilometre area of open common-land owned by Bradford Metropolitan District Council. It lies on the north-west boundary of Baildon village, dominating the towns of Bingley and Shipley which straddle the River Aire in the valley below. Today the moor is ungrazed and used almost exclusively for recreational activities. Much of the area is reverting to bracken and tree cover with the exception of managed areas, such as the golf course and numerous track-ways.

The basic rounded, stepped, profile of the landscape has changed little since the retreat of the last phase of the Devensian glaciation from the area, approximately 12,000 years ago. Initially the landscape was relatively inhospitable, but as time progressed and the climate warmed, trees and shrubs started to clothe the bare hillsides above the wet, marshy valley bottoms. Wildlife began to provide a reliable food source for people moving across the region around 9-10,000 years ago. Evidence for these early people comes from several small collections of stone tools recovered from the valley bottom and lower slopes.

It is possible that mineral resources were exploited from earliest times, expanding and developing as technologies evolved and industries began to be established. During the twelfth and thirteenth centuries Cistercian monks were involved in the production of iron in Yorkshire. Records show that Rievaulx Abbey, near Helmsley in North Yorkshire, held lands at nearby Harden and Sconce, and a Grange at Faweather, with the latter two sites being on the eastern edge of Baildon Moor. Early twentieth

Baildon Moor in the 1960s. Remains of pits appear in the background as large potholes

century Ordnance Survey maps of the area indicate 'bloomeries' alongside Loadpit and Glovershaw Becks.

It is 1387 before we find a definitive written record of mineral extraction in the area. At the time, the Lord of one of the Baildon manors, John Vavasour, complained that several people were digging coal on his land, which resulted in lost revenue of one hundred shillings. Saxton's map of 1610 refers to the 'cole pittes on Bayldon Common', while seventeenth and eighteenth century records are littered with references to disputes, sales and leases of mineral rights. The earliest mid-nineteenth century Ordnance Survey maps of the area record many hundreds of 'Old Coal Pits'.

The picture today is one of an area covered with the surface evidence of centuries, if not millennia, of exploitation and small-scale industrial activity, and we now know that coal, fireclay, ganister and ironstone were mined in significant quantities, in addition to the extraction of building and road stone, sand, and pottery clays.

Background to the survey

In 1900, following the 1898 sale of the Moor to Bradford Corporation by William Wade Maude for £7,000 and retention of the stone extraction rights, concerns for the preservation and long-term security of the many prehistoric antiquities on Baildon Moor were expressed by members

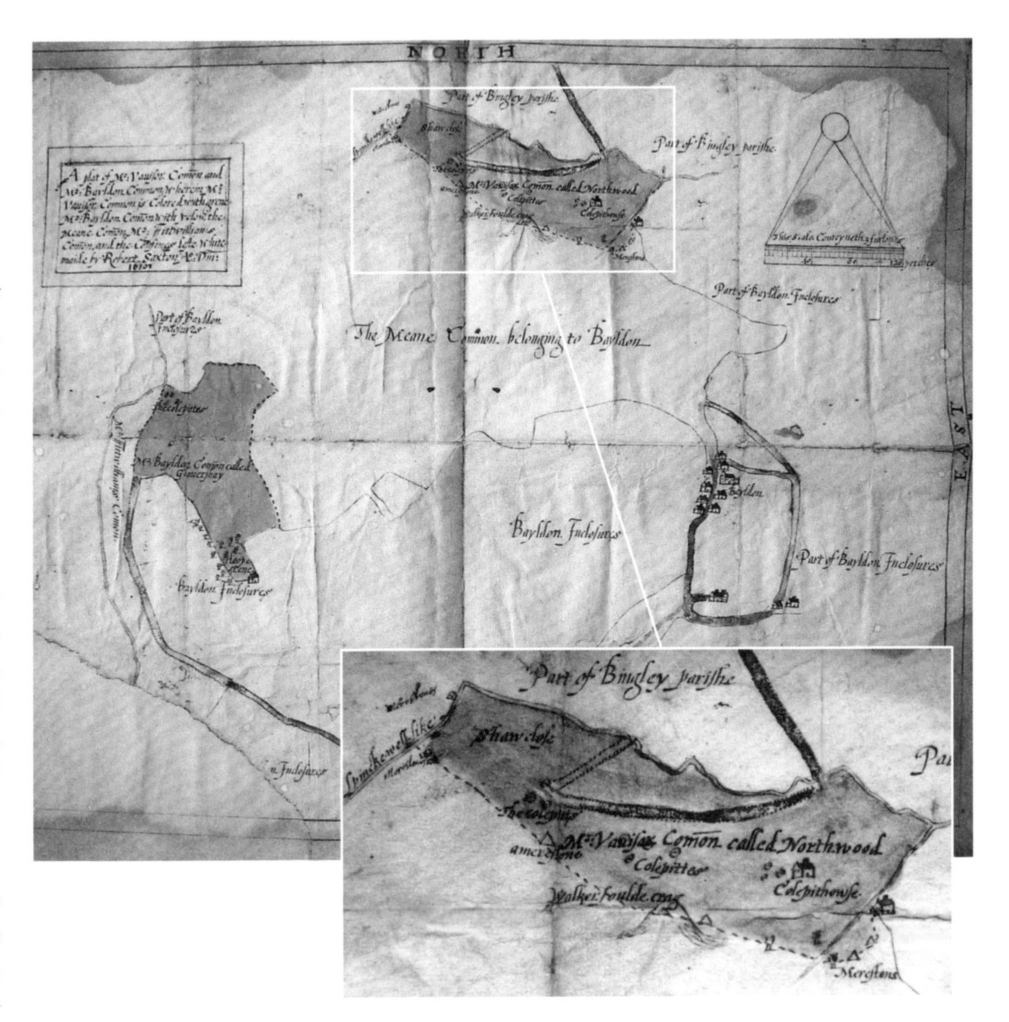

Saxton's map of 1610 showing Baildon Moor and the 'Colepittes'

of the Bradford Historical and Antiquarian Society. The Society subsequently undertook a detailed survey of the sites. Just over a century later a group of local residents began to have similar concerns, but for all sites, not just those of prehistoric origin. In 2009 the Friends of Baildon Moor formed with two of their constitutional objectives being 'to preserve Baildon Moor as an open space amenity for education' and '...to promote the protection of the physical and natural environment...'. The Friends, and other local people, wanted to develop a deeper historical understanding of the Moor's natural resources, its relevance to local people and the socio-economic development of Baildon and neighbouring towns. An opportunity to fulfil these objectives and interests was presented by the Riches of the Earth Project. A launch event was held in September 2010 and from this a group of representatives from Friends of Baildon Moor, Baildon History Society, the Stanbury Hill Project and other individuals was formed. Individuals' interests, experience and desires meant that some members of the team preferred historical research, while others preferred to undertake on-site survey and recording. It was quickly realised that the extent of the area and number of sites to survey at the detail we required would result in many months, if not years, of work.

The survey element comprised three phases: detailed measured survey of a selection of circular negative features; GPS (Global Positioning System) survey of three groups of features; and a landscape-style survey of one area of Baildon Moor. To carry out this phase of the survey, the team chose a methodology that had been developed by members of the Watershed Landscape Project's 'CSI (Carved Stone Investigation): Rombalds Moor'. This simple, fast survey method suited our limited resources. Using 100m^2 images from Google Earth tied into the Ordnance Survey grid using a hand-held GPS unit, we plotted our results onto transparent drafting film laid over the images. During the survey each square was identified using GPS and ranging poles were placed at each square's corners. All the visible pits, paths, and other features were marked in pencil on the overlaid film. In addition to the master drawings, standard data recording sheets were developed using experience gained on other survey projects. These gave consistent data recording. Completed overlays have been stitched together using Geographic Information System (GIS) software to create a master map that covers the whole survey site.

Evidence of coal mining on Baildon Moor

A walkover survey revealed numerous roughly conical pits of variable size. These indicated capped and partially collapsed vertical mine shafts. We found these pits either dry or filled with water and surrounded by raised banks of grey mine waste. Some of the pits were found with platforms of the same waste material. The overall arrangement of the pits could be seen clearly from aerial photography.

The first coal to be extracted from the moor would have been obtained through simply digging into surface outcrops. This technique would then be expanded into short tunnels or drifts and then shafts as the seam extended back into the hillsides. 'Old pits' are marked on the first edition 'England - Yorkshire: 201', Ordnance Survey 1:10560, 1852, and they have all, if somewhat mistakenly, been labelled as 'bell pits'. Bell pits were rarely deeper than 10m and whilst their construction and operation was simplistic they weren't without danger. They were formed by sinking a shaft down to a horizontal coal deposit which was then worked in all directions until lack of roof support or ventilation made further mining unsafe. Coal was lifted to the surface by basket, using a winch, or perhaps on

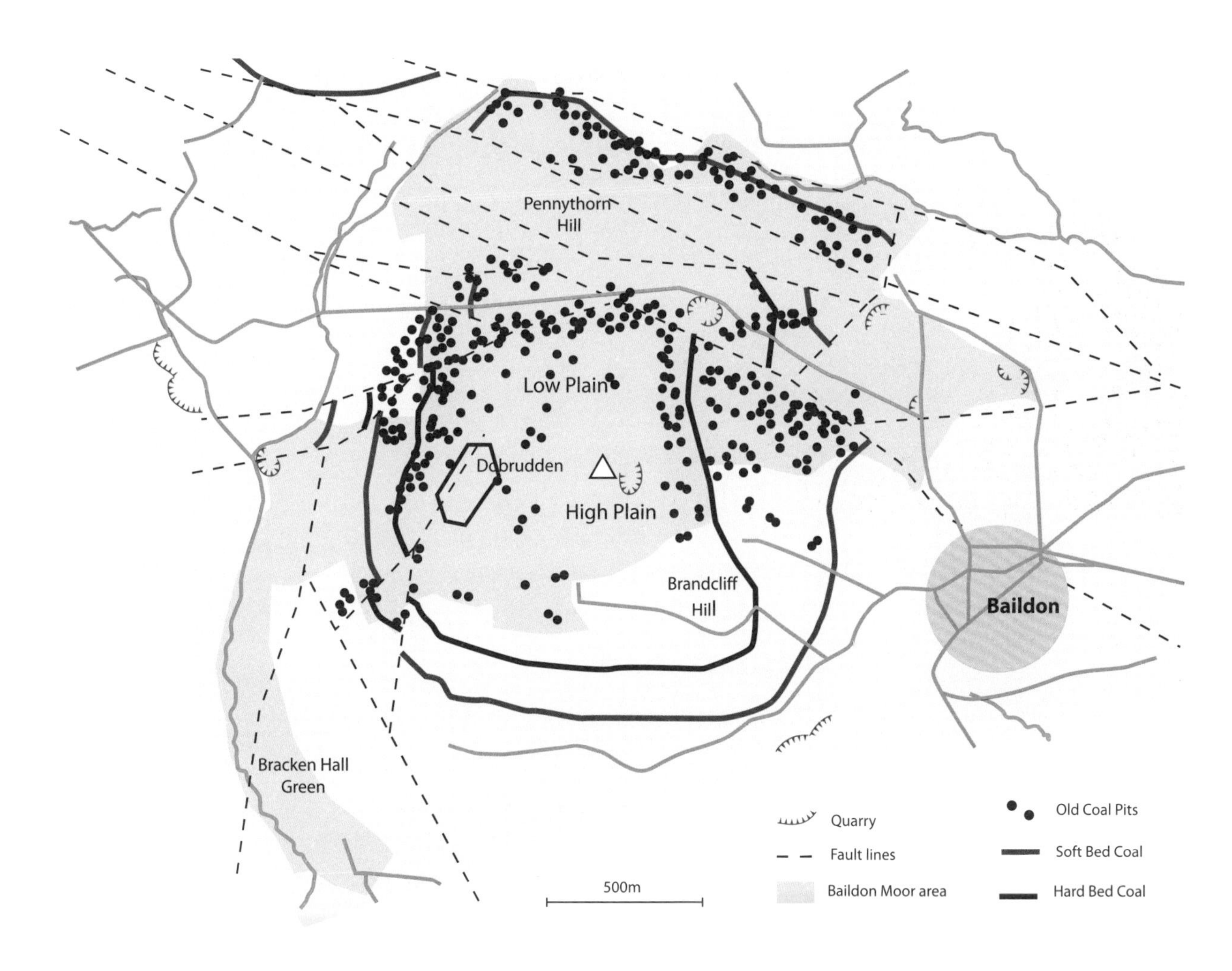

Map showing the main coal seams and the location of coal pits (as they appear on the first edition 'England - Yorkshire: 201', Ordnance Survey map 1:10,560, 1852). Based on the BGS Geological 1:50,000 Sheet 69 Bradford (Solid and Drift Edition)

Members of the Baildon Moor team surveying one of the many circular depressions across the moor

the larger and later sites with a horse and gin. Horse gins were still in use in the Bradford and Halifax area in 1841. On abandonment any unusable material was deposited back into the shaft, leaving the depressions in the landscape that we see today.

As the shallower seams became depleted, the need to mine deeper became necessary. A large number of the Baildon Moor shafts gave access to the deeper soft bed seam which was exploited through a system of galleries. This technique, referred to as pillar-and-stall, favoured workings below the economic bell pit depths of 10m but not much greater than 250m. Shafts would be dug down to the seam level which was then worked horizontally along roads or galleries leaving large pillars of coal. The pillars would then be extracted in a controlled manner and the roof allowed to collapse. Once the mines had grown to this scale and were becoming a major industrial activity the problems of access, coal removal, ventilation and drainage became major obstacles. These engineering and economic factors were frequently the reasons for the closure of the workings.

We also found other features such as moor track-ways, which were also marked on the first edition 'England - Yorkshire: 201', Ordnance Survey 1:10560, 1852.

These might reflect earlier industrial activity, although some track-ways are seemingly modern in origin. The moor was also the location of rifle ranges through the Victorian period and was used for tank training during World War II. These too have made their mark on the landscape.

Documentary evidence

The Baildon Moor pits were valuable assets which were fought over in court. They were set up in the early seventeenth century by a William Baildon, together with Gervaise Fitzwilliam, the owner of Hawksworth Manor. After William Baildon's death his estate, including the pits on the moor, were managed by the trustees of the estate until the six year old Francis Baildon, grandson of William, came of age. By this time the whole estate was in a real financial mess and in 1652 Francis sued Sir Richard, his father-in-law and one of the estate trustees. The court ordered that the two parties should share the mining revenues in future, and Francis was paid £100 compensation.

Documents also tell us about the challenges of working a mine. A 1751 lease between the Lord of the Manor, and Ann Butler of Baildon, examined at West Yorkshire Archives Service, Bradford, indicated that a drainage problem was a concern for the mines.

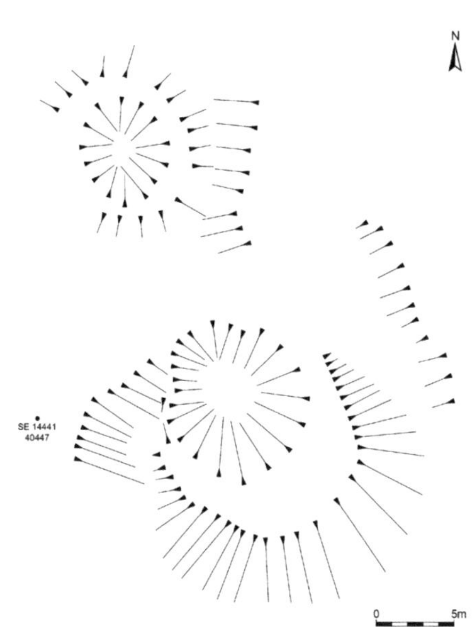

(above) Detailed survey of two of the circular depressions, based on an original plan drawn by members of the survey team

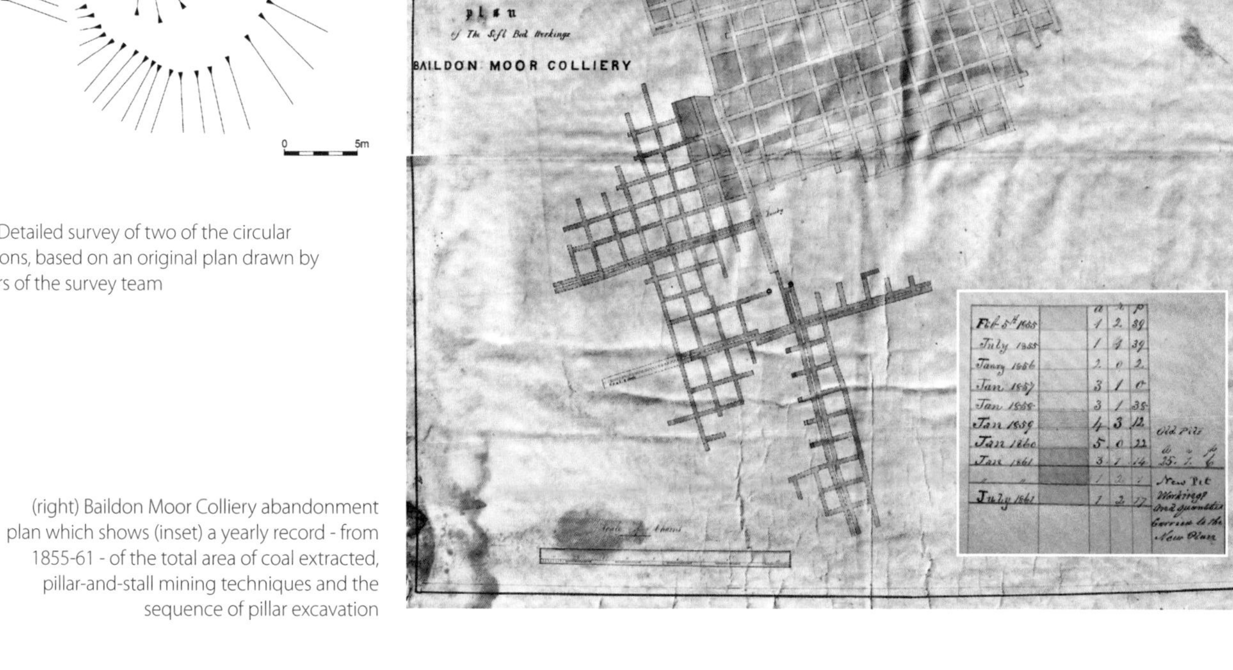

(right) Baildon Moor Colliery abandonment plan which shows (inset) a yearly record - from 1855-61 - of the total area of coal extracted, pillar-and-stall mining techniques and the sequence of pillar excavation

Safety was also a concern. The Royal Commission report on Children in the Mines, published in 1842 paints a picture of the harsh conditions endured in the local mines. Returns on the deaths from accidents and explosions in the Bradford and Halifax district show that between December 1837 and June 1841 there were 50 recorded deaths in these areas. One death, which occurred on 22 April 1838, was reported in a local newspaper. Joseph Craven, aged between six and seven years old, fell from a corve (basket) on his way down a shaft, falling some 70 yards (approximately 65 metres). Joseph was working for Baildon Moor Colliery.

For most of the long history of coal extraction on the moor, the industry was dominated by small-scale, probably under-capitalised concerns. At any one time there would have been several areas being mined. In addition the Lord of the Manor's local agent would have observed all the operations. William Midgley was perhaps the first businessman to have several mines working at once. Midgley was a very prominent person in Baildon village. He did many things in his life, starting as post master for Baildon. He was an original member of the Baildon Local Board of Guardians when this was established in 1852 and was appointed as Chairman of the Board in 1854. Various partnerships worked the mines - Ellis, Morrell and Beck, and later Midgley, Beck and Co., with William Midgley eventually owning all the larger mines. He also built houses for his workers and a larger house for himself which still stand as East Parade in the centre of Baildon.

Little mining activity has taken place since the sale of the moor to Bradford Corporation, although some pits may have been opened up during the depression years of the early twentieth century.

In addition to coal, ganister was also extracted by the Yorkshire Ganister Company, who mixed it with clays from the moor and from near their brickworks at Baildon Green. Stone was also extensively quarried. Early Ordnance Survey maps name several delphs on and around the moor.

Prominent features of the moor are the large spoil tips, known locally as 'The Cinder Caves'. Spoil tips from older coal mines are prone to spontaneous combustion because of their high coal content and therefore their combustible gas content. They can easily be set aflame by sparks generated by landslips on the spoil tip. Burning the spoil heaps can also make the shales within the tip hard enough to be used for road stone. That is probably why these spoil tips were deliberately burned.

Oral history evidence

Michael Dwyer, a current Baildon resident, recorded information about his father's and grandfather's recollections of the last pits on Baildon Moor. There were three pits, at Brancliffe, Dobrudden and Lobley Gate, and they were all worked by William Midgley. A 17 year old Matthew Dwyer, Michael's grandfather, came to Baildon to help with problems at Lobley Gate around 1890. Despite his young age he had wide experience, having already worked as a trapper and hurrier moving tubs from the coal face to the surface, and finally as a hewer at the coal face. He lodged at one of Midgley's houses with a Ben Jowett. He later married Ben's daughter, Ada.

After the closure of Lobley Gate pit, Matthew is said to have continued to work in bell pits extracting coal, ganister and pottery clay. Michael, as a child, was shown the various bell pits his grandfather had worked.

On the back of a photograph, Michael's father wrote:

> *'My grandfather, Ben Jowett (1838-1913), used to be deputy at the coal pits on Baildon Moor. I don't think there was much coal,*

Ben Jowett, a deputy at Baildon Moor coal pits

and it was only a meagre affair and as far as I know there was only one [pit] there, when I was a boy. There was a tunnel going underground on the eastern side at the bottom of the moor top, with narrow gauge rails, about a yard or so wide, running into the bottom of the hillside, with one small bogey truck.'

Much of the oral history has been confirmed by searching local newspapers (1876-98) and Baildon Local Board Council minutes from 1852. These give contemporary accounts of some of the pits and the activities associated with them.

Baildon miners in the census records: 1841-1901

A survey of the censuses showed a pattern of people working in coal extraction:

1841	*8 men*	*3 boys (defined as under 15)*
1851	*0*	
1861	*42 men*	*10 boys*
1871	*21 men*	*7 boys*
1881	*16 men*	*1 boy*
1891	*4 men*	
1901	*0*	

In 1841 a few families were working on a small scale, perhaps obtaining poor quality coal from a relatively easily-accessed seam. This became uneconomic or maybe the landowner objected. In 1851, whilst there were no miners, there were three coal dealers, presumably importing coal for use in the mills. Then in the mid-1850s William Midgley and John Beck invested in large-scale modern pits extracting from a deeper seam. They brought in skilled men from elsewhere.

A more detailed look at the 1861 miners showed 25 born in Baildon, Shipley or Bingley, 17 born fairly locally and 10 incomers.

Many were sons of miners and had learned their skills from their fathers. Many 'fairly local' miners had been born in Thornton or lived there. In 1841 or 1851 several men had worked in the Ingleton area, where mining continued until 1936. More surprising to us, was the presence of 'incomers' from Dentdale and Barbon, near Kirby Lonsdale.

Many parts of northern England had limited supplies of coal, often of indifferent quality or in narrow seams which could only be worked on a small scale.

After 1861 improved transport and economies of scale from large-scale mining meant that local coal production was no longer competitive or profitable.

Conclusion

Throughout the past year the team has surveyed approximately 0.4 square kilometres, which represents over 12% of the moor's total area, through a combination of techniques, in addition to more targeted work prior to this. We have spent many dozens - if not hundreds - of hours undertaking research involving investigation of historical documents, literature, papers and a multitude of records, in an attempt to gain further insight into how the people manipulated and utilised the natural resources provided by the landscape and used it to their advantage.

Despite what we have learned questions remain. Are the 'old pits' marked on early Ordnance Survey maps attempts to plot individual pits accurately, or simply formal cartographic illustrations? By comparing the overall arrangement of extraction pits with known illustrations of shaft mining elsewhere can we understand how the Baildon pits were dug, operated and perhaps drained? Coal was probably lifted out in corves by human or horse power, and wooden gins, although no evidence of any timber shaft superstructure remains. Many of the pits must represent ventilation shafts sunk to remove explosive gases. Drainage must also have presented a problem to the miners. Were underground channels called 'soughs' used to drain the entire system of galleries, or were shafts sunk below the coal seam to create sumps in which water would collect?

Though we have a great deal of information about the exact location and dimensions of the pits, we have not yet reached a point where we can firmly conclude what each pit was for or how it was operated. Being able to do this would lead us to likely date ranges for the pits, and to understand how extraction was carried out and developed over the centuries.

The project's greatest success to date has undoubtedly been in bringing together individuals from various community organisations in Baildon, such as the Friends of Baildon Moor and Baildon History Society, and other individuals interested in industrial archaeology, to pool their knowledge and experience. Through training the project has provided volunteers with skills in surveying techniques, explaining the geology of the area, understanding how to read the landscape, and how to use handheld GPS units. We now have a strong nucleus of dedicated individuals with the skills and enthusiasm to continue to move forward and work towards unravelling the answers to the many questions. Documentation from archival sources, maps and plans, photographs, evidence from oral history have all been gathered together and will provide a valuable resource for the community.

One of the earliest known images of Baildon, taken at a feast in the 1880s. Local people are posing in front of a smithy where corves for the mines were made

Survey 2: Oxenhope Moor, A Past Written in Stone

A survey was carried out of the extraction features of an area of Nab Hill on Oxenhope Moor often known as 'on t'Nab' or 'on t'Fly'. Quarrying was intermittently carried out here several times during the twentieth century. Although an earlier quarry was present on Warley Moor in 1709, most of the quarries in the area date from a period of expansion in the 1840s to the decline of quarrying in the area around 1900.

Background to the project

Nab Hill Delphs are a series of disused sandstone workings set on the prominent escarpment and plateau of Nab Hill above the village of Oxenhope, near Keighley, West Yorkshire. Historically the small quarries, or delphs, extended south along the escarpment edge into the neighbouring Township of Warley, in the present Metropolitan District of Calderdale. This upland area slowly evolved as waste land was enclosed through the seventeenth to early nineteenth centuries, until a hamlet known as Fly was established.

Although consisting of separate farmsteads, the residents must have considered themselves a community with close ties to their neighbours in this relatively isolated location. A plan taken from the archives of the extensive Castle Carr Estate held at West Yorkshire Archives Service, Calderdale (see page 31), shows a typical upland settlement pattern of areas of marginal farming supplemented by domestic textile production.

For the purpose of the study of quarrying in the area, the hamlet is considered to have extended from Nab and Nab Water farms in the north to Knoll in the south, although census returns

Large cellar tops covering coal cellars under a street in Haworth. These were typical products from quarries such as Nab Hill

show us that many of the workers lived further afield. Whilst early quarrying may have been carried out in the south of the area in Area 4, most of the workings date from the expansion of quarrying from 1840 onwards. Some enterprises were of a commercial nature controlled by merchants. Those quarries in Area 4 saw the migration of skilled workers from the important quarrying area of Extwistle near Burnley. Other workers were opportunists exploiting the accessible flagrock and providing a supplementary income to local farmers following the decline of domestic textile production. Due to their isolated position the workings never evolved into large-scale mechanised concerns

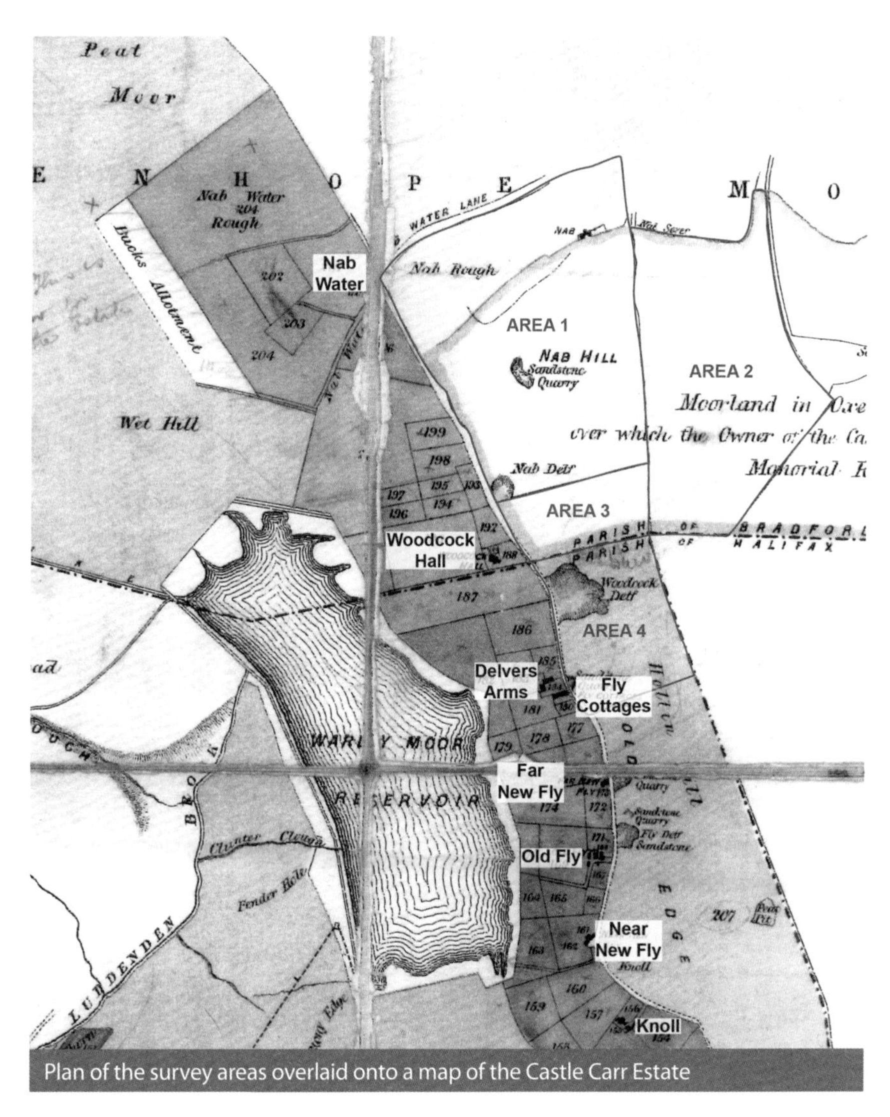

Plan of the survey areas overlaid onto a map of the Castle Carr Estate

and they continued to be worked by hand until their decline around the beginning of the twentieth century. Later mechanised quarrying destroyed much of the early workings but those remaining can be considered good examples of hand extraction of flagrock of the mid- to late-nineteenth century.

Aims of the project

The stone extraction industries have received little attention in the field of industrial archaeology, and so the aim of this project was to investigate and record a quarry site. This was achieved by a programme of documentary research and fieldwork. As well as teaching community volunteers the basics of topographical survey it was hoped we would be able to present our findings in a metrically accurate plan of the site. In order to understand the area better an initial appraisal of the quarrying remains on Oxenhope Moor was completed. This considered both the historical development of the settlement and exploitation and also how the underlying geology dictated extraction methods. In light of this it was felt that the Oxenhope Moor remains were sufficient to allow us to test theories about methods of hand extraction of considerable amounts of building stone.

Documentary research

Aerial photographs revealed a complex series of old quarry workings scattered along the north edge and eastern side of the Nab Hill plateau. To the south large areas of more recent workings seemed to have obliterated any evidence of earlier working. Documentary research carried out in various local archive collections began to piece together a picture of complex ownership and tenancy patterns that allowed us to better understand what had appeared to be random workings. This, coupled with study of the geology and a consideration of known quarrying techniques through oral history, allowed us to conclude that the workings on the north end of the plateau were typical of the delphs of the period between 1840 and 1900. As many of these have been lost to later working or household tipping it was decided to record one select area that was identified on an estate plan from 1880 and on historical Ordnance Survey maps in the archives.

A plan of Nab Farm, identified as Area 1 in the study, showed five separate leased delphs which probably reflects the height of working around 1880. Rate Book records revealed who was occupying the delphs in 1865 when the farm belonged to James Ratcliffe. At that time only two quarries were being worked, one by Thomas Binns and Jacob Ellis, and the other by Robert Sunderland. It can be assumed that one of these is the delph that we surveyed. This would seem to indicate that the real expansion of the quarries is associated with large-scale reservoir construction of Warley Moor (Fly Flats) Reservoir for Halifax, and Thornton Moor Reservoir for Bradford, along with their extensive water catchment schemes which occurred over the next two decades.

Survey methodology

We decided to survey the chosen delph by hand using the tape-and-offset method at quite a large scale. In this instance a scale of 1:200 was chosen for two important reasons. Firstly,

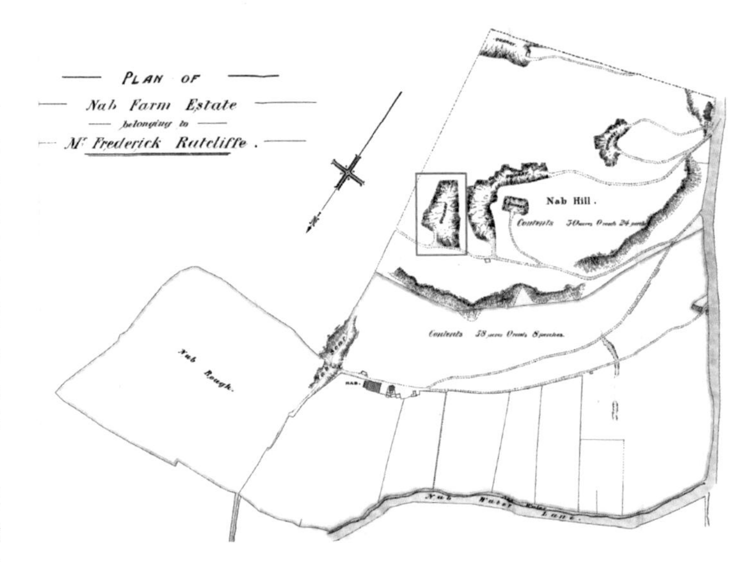

Nab Farm Estate. Delph survey area highlighted by red rectangle

our project was a community archaeology one that allowed the training of local volunteers. The progressive process of plotting a site allowed volunteers to quickly become competent in carrying out metric surveys and importantly to understand that process. Secondly hand survey of complex sites allowed close interpretation and analysis throughout the recording process that is often not possible through more technologically dependant processes that are by their very nature 'remote'. The fact that hand survey can be carried out without resort to expensive electronic instruments again makes it ideal to teach to community groups operating on small budgets.

Photographic survey techniques were used to fully illustrate the site both to provide a record of the site and hopefully to allow

HIGHWAY RATE. DISTRICT OF HAWORTH TOWNSHIP —— LOCAL BOARD

RATE.

Names of Occupiers, or Person Rated. 2	Description of the Premises or Property Rated. 3	Situation of Property.	Owners.	Estimated Extent.			Annual Value. 4		
				A.	R.	P.	£	s.	d.
Ezra Stirk	House &c	Nab End	James Ratcliffe Esq				1	15	-
John ~~Hitchen~~ Ratcliffe	"	"	"				1	15	"
Thos. Binns & Jacob Ellis	Stone Quarry	"	"				11	5	"
Robert Sunderland	"	"	"				7	10	"
William Dawson	"	Woodcock Hall	J. P. Edwards Esq				11	5	"
"	House &c	"	"				2	10	"
John Dawson	House, Barn &c	"	"				2	14	"
"	Land	"	"	101	2	"	17	13	"
William Ferrand Esq	House, Allotments &c	Keeper's Lodge	"	172	"	22	8	12	"
Henry Greenwood	House, Barn &c	Nab Water	"				2	18	"
"	Land	"	"	55	1	4	13	6	"
	House &c	"	"				1	15	"
	Allotments	Bentley Allotments	"	106	3	26	5	"	"
Jonathan Whitaker	Plantation	Nab-Scar	Hemmingway	4	1	4	"	8	"
				440	"	16	88	6	"

Extract from a Haworth Rates Book

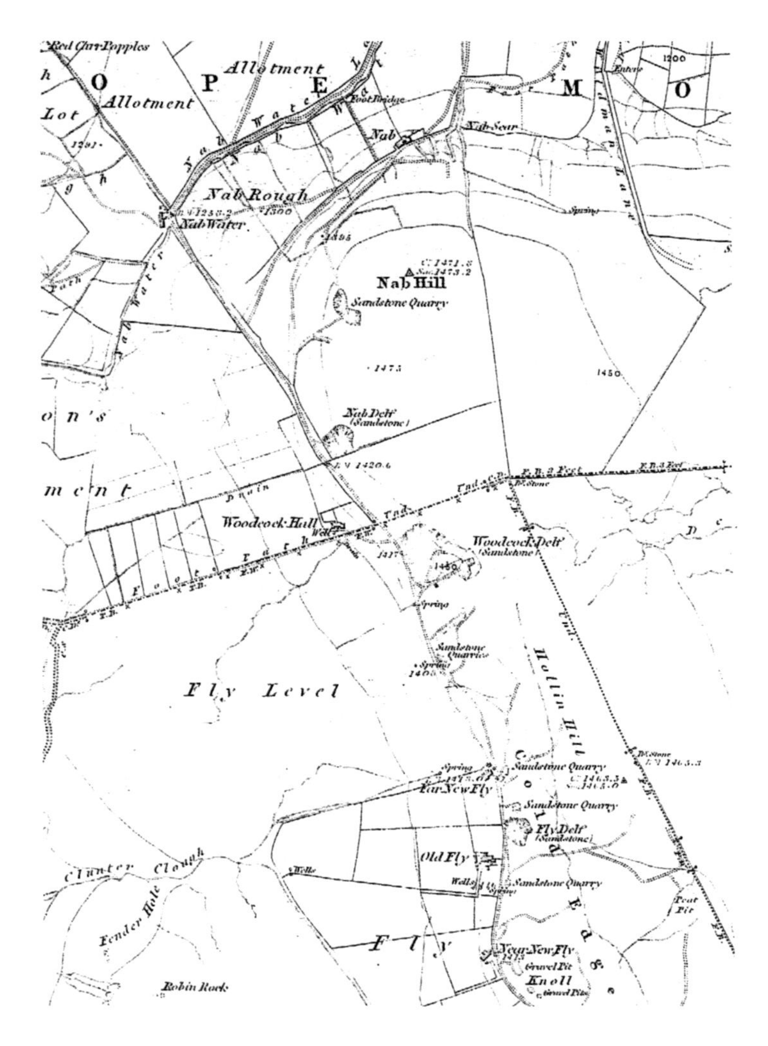

Detail from the first edition 'England - Yorkshire: 215', Ordnance Survey map 1:10,560, 1852

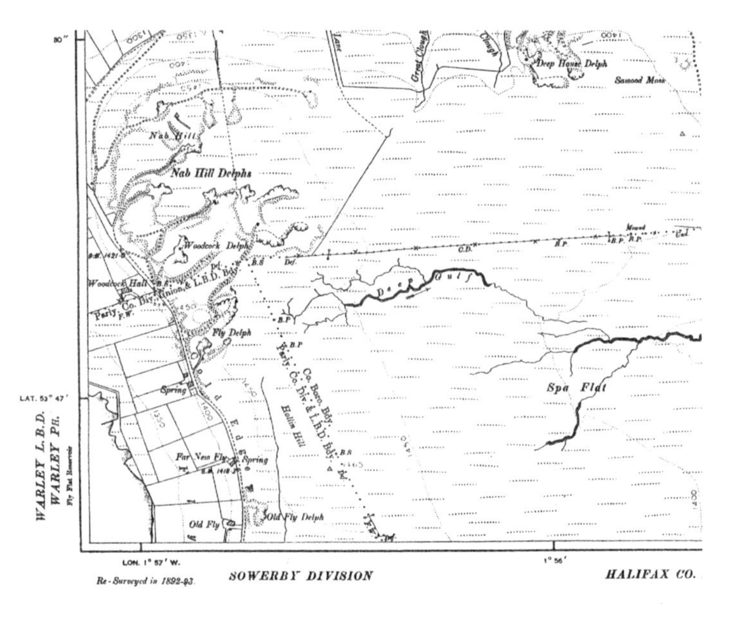

Detail from the second edition Ordnance Survey map 1:10,560, 1894

further interpretation and analysis. The findings were presented in a series of illustrations including phasing and suggested extraction methods.

In order to aid interpretation and to clarify understanding, one volunteer with a graphics and archaeological background was tasked with producing a reconstruction drawing of the delph as it may have appeared mid-to-late nineteenth century (right).

Reconstruction of quarrying on Nab Hill

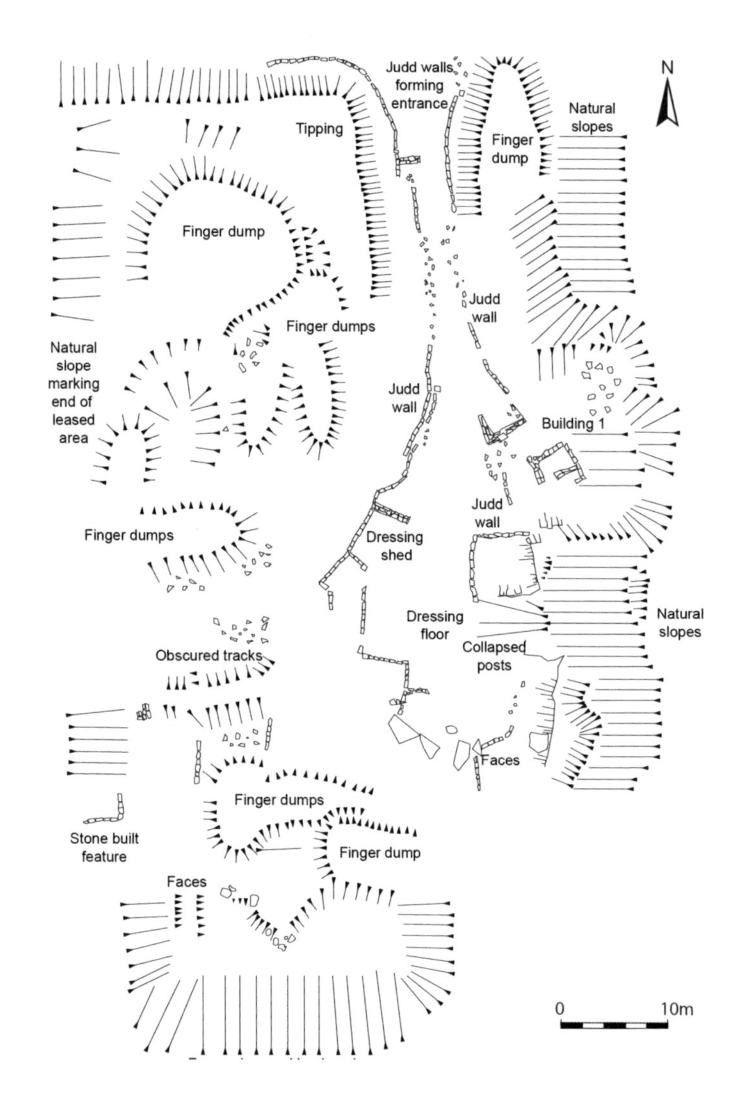

Plan of the Nab Hill quarry, based on an original plan by the Oxenhope survey team

Findings

Close observation of the site, made possible by manual recording at a large scale, suggested that a considerable amount of evidence remained of the extraction methods of a small flagstone delph of the mid- to late-nineteenth century. The survey revealed that the site corresponded to the full extent suggested by the historical plan from around 1880. It also revealed that much of the western side already quarried had been tipped over with overburden cleared to expose the next area of the delph. The eastern half remained as a large area where it is suggested the flagrock had been extracted and the site levelled. This levelled area was interpreted as a dressing shed and dressing floor. In local quarries large lumps of flagrock are lifted clear of the working face and processed on a dressing floor by splitting, or 'riving' the rock into handleable sheets. These are then further processed by dressing, or 'fettling', into squared flags in an open-fronted dressing shed. Down the eastern and along the southern boundary of the delph there were rectangular areas of natural rock left aside to be 'got' later. This is consistent with the systematic removal of flagrock in small areas, 'posts' of stone, formed by vertical seams. This part of the delph is flat, covered with turf and could conceal a series of 'judd' retaining walls that could further confirm the suggested methods of extraction.

Members of the Oxenhope team surveying the Nab Hill quarry

Survey 3: Changing Views on Todmorden Moor

Recording Todmorden Moor's industrial heritage

Todmorden Moor lies on the Pennine watershed and abuts the border between Lancashire and West Yorkshire.

It is known locally that Todmorden Moor has been the site of industrial activity for many centuries, although most activity had come to an end by the 1960s leaving parts of the moor with a reputation of being a 'degraded landscape'.

The Todmorden Moor Restoration Trust was formed in 1992. The members of the Trust and other volunteers have worked hard since then to transform the moor from a rubbish tip to a pleasant place to walk, run, ride bikes, enjoy moorland wildlife, and to 'take the air'. Some areas of the moor proved difficult to improve however; spoil heaps from abandoned coal mines are not pretty and are impossible to hide.

From 2007 members of the Trust were unhappy about the possibility of a wind farm being constructed on the moor. The Trust was concerned about the impact that the wind farm construction might have, amongst other things, on local water supplies that percolate through the peat, underground workings and sandstone. As a result much research was done into the underlying geology of the moor.

A local retired miner gave us access to a number of maps which threw light on both the geology of the area and the associated industrial activity. An extract from a geological survey of Great Britain undertaken in 1844 clearly shows the Coal Measures that attracted miners to the moor.

Abandonment plans show us that much of the moorland overlies a honeycomb of mines. It is sometimes possible to link the signs of collapse on the surface with records of underlying workings. Many surface features, however, appear not to be directly related and these indicate mining and quarrying activity on a small scale to meet the immediate needs of local people.

It quickly became apparent that the area has a complex underlying geology and that the moor had been exploited for the mining of coal and clay for hundreds of years. Of particular interest was the discovery that the moor had become well known in Victorian times where the Union Seam of coal contained rare 'coal balls' (see page 10).

These coal balls attracted a visit from the British Association in 1903. The Todmorden and Hebden Bridge Advertiser recorded that 'the principal object of the visit was to study the limestone nodules which are a remarkable feature of the coal strata in Dulesgate and are rich in fossil remains . . .'. The party, which included 'half a dozen ladies' who had to remove their hats, reached the coal workings 120 yards (approximately 100m) below the surface via a '1241 yard long tunnel from six to eight feet high and the same dimensions wide'. There was a 'special train of about 20 coal tubs provided to transport the company underground'. These workings were known then to have begun in 1855 but many abandoned old workings are also mentioned.

As a result of our coal ball research, West Yorkshire Geology Trust (WYGT) became involved and the area was designated as a site of geological interest. The South Pennines Watershed

Extract from the Geological Survey of Great Britain, surveyed in 1844-7 by Captains Tucker and Berry. The survey clearly shows the coal measures which have attracted miners for centuries

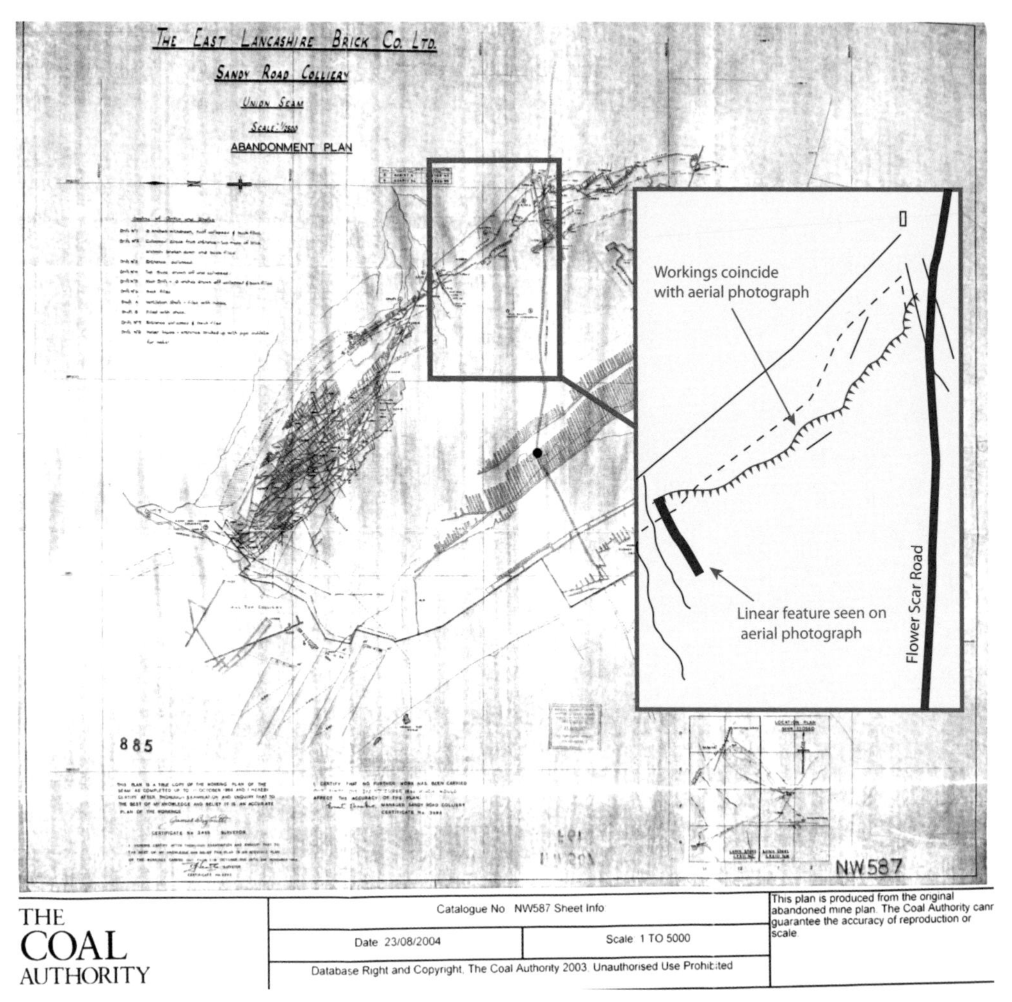

East Lancashire Brick Co. Ltd Sandy Road Colliery Union Seam Abandonment Plan. This shows coal workings underlying one of the areas recorded by the team. The seam was worked between 11 October 1844 and 12 November 1953

Landscape Project saw the need to record the signs on the moor where past industries have extracted coal, clay and stone.º For Todmorden there was some risk that much of this would be destroyed by the construction of the wind farm. A Geology and Heritage Trail was subsequently developed (see page 58) across the moor with the help of the WYGT, highlighting the geology and resulting industrial archaeology still visible on the moor.

Our small group of volunteers was trained in simple archaeological survey techniques, following the same landscape-style methodology employed on Baildon Moor, and also in recording oral histories from people who had worked in industries associated with the moor. Soon we had a team with the required tape measures, poles, hand-held GPS units, cameras, clipboards and recording sheets and some knowledge of how to use them. In February 2012 we had our first attempt at using our new-found skills.

The moor was divided into a total of 325 squares, each 100m x 100m, to be surveyed using the techniques of

The team measuring one of the many depressions on the moor. (inset) A conical depression recorded by the team

measuring and recording in which we had been trained.

This was clearly going to be a major challenge. We found that a team of three or four people could survey one square on the ground in about two hours, depending on how many features were within the square. In addition to the fieldwork, at least as much time is spent sorting photographs and hand-drawing diagrams of the features.

Some of our team had full-time jobs so were limited to evening and weekend field work. We also had the threat of imminent wind farm construction but luckily one of our team was able to suggest a faster way to proceed. He is a member of a model aircraft flying club that regularly uses Todmorden Moor. He has developed a model aeroplane able to take aerial photographs using a digital camera which also records the GPS positions. Using these techniques, we could take aerial photographs of the whole of the moor at much higher resolution than available on Google Earth.

By careful examination of the photographs it was possible to prioritise the 100 metre squares which contained interesting features. We could also identify squares which had no visible features worthy of investigation on the ground. We selected around 30 squares which have significant features requiring a full survey.

In the areas chosen for further examination, there is a wealth of evidence for human activity, often in the form of troughs, conical depressions, mine entrances, half-buried walls and occasionally deep shafts which required us to be careful where we put our feet. On one memorable occasion a call to the Coal Authority was needed so that they could fence off dangerous collapsed workings. This was on a day that we were surveying the area where coal and clay were extracted to supply the ceramic pipe industry on the edge of the moor (see page 43).

As we became more adept at finding features which had shown up in the aerial photos, members of the team became increasingly interested in the work. What had previously been seen as a grassy mound or a muddy ditch was revealed as evidence of a landscape that has been managed and used for hundreds, perhaps thousands of years. We began to see how the story of the old mines shows itself across the modern-day landscape and we could imagine the many hundreds of people who have worked on and under the moor over the centuries, often in terrible, dangerous conditions.

We have now completed almost a year of sporadic surveying on Todmorden Moor - sporadic because we had possibly the worst year of weather in which to work. So did the team learn anything? We are sure we did.

It is no secret that Todmorden Moor has been heavily exploited for minerals in the past. Parts of the moor appear as just a frantic muddle of holes and spoil heaps. We have begun to make sense of that confusion.

The steep gradient of the coal seams of 1-in-7 or 1-in-8 and frequent faulting under the moor made mining difficult. We now understand that the extent and confusion of mining evidence on the surface indicates many different attempts to extract coal over the centuries as new drifts were dug and old ones were abandoned or collapsed. But we could not say how early the first mining took place on Todmorden Moor and whether some of the conical depressions we surveyed are early bell pits. That would be a good thing to establish one way or the other.

One of many holes which we found and measured overlying the area of clay workings. The clay was used to supply a factory on the edge of the moor which manufactured ceramic land drains and sewerage pipes

Aerial photograph showing the area of collapse along a line of the clay workings stretching across the moor

The aerial photograph clearly shows a line of depressions adjacent to a track associated with the coal outcrop. As these depressions are an indication of a collapse, survey was not carried out here as to do so could prove dangerous. A linear feature can also be seen on the photograph, leading from the track to the line of depressions. This was later discovered to be a cutting for a tramway into a short-lived drift mine

Photograph showing the linear feature (centre) as it appeared on the day we made the sketch below

Photograph taken in the 1960s whilst the mine workings were still visible

89891, 25000 linear feature extending to 89880, 24969 (N.E. to S.W.) Evidence of worked stone along W. side. No stone visible to E. side, although the turf and moss were not springy to walk along. The cut is straight, not meandering, and c. 5m. wide. A possible adit.

N.

Isolated blocks of stone following the line of the feature

line of unyielding surface

An extract from a recording form which shows the linear feature visible in the photographs

John Tommy Earnshaw worked most of his life in the pits on Todmorden Moor. He is standing at the entrance to one of the coal mines in this area

Temperley Brothers' ceramic pipe works in the 1940s. Temperleys sold out in 1961 and by the mid 1970s the site was semi-derelict. Very little now remains

Thanks to the Coal Authority's abandonment plans and an excellent book *Industrial Heritage - A Guide to the Industrial Archaeology of Bacup and Stacksteads*, written by Mike Rothwell in 2010, we can now match up some of this information with surveyed evidence still visible on the ground.

People with local knowledge tell us that some mining machinery and small pieces of equipment were simply abandoned underground when the last mines closed. Of possible interest to the archaeologists of the future? Only a very disturbed landscape remains but we could still identify and record what must be some of the first shallow drifts from 'day-eyes' along the coal outcrop as the first miners followed the coal seam underground. We recorded evidence of mine entrances, mineral tramways, possible air shafts, conical depressions in lines across the moor, and some shaft collapses into deep workings below.

Our next job is to match up our survey work with existing historical records. At least we can do that work in the dry!

Heritage Trails

Take Moor Care

Please take care when walking in the uplands. The weather can change quickly, so ensure you have suitable shoes, clothing, a map, and food and drink. Whilst exploring the landscape, always follow the Countryside and Moorland Visitors Code. Be vigilant about avoiding wildfires and report any fires to the police or fire services. Keep dogs on short leads, especially during ground-nesting bird season, and please leave the uplands as you found them.

The Baildon, Oxenhope and Todmorden Moor landscapes are covered with remains of a bygone age of mining and extraction industries. Old mines and industrial works can be extremely dangerous. Never attempt to go inside or explore the remains. If you notice any areas of potential danger, to yourself or others, please notify the appropriate authority.

Coal Authority

24-hour emergency call out service: 01623 646 333 for issues relating to public safety including collapses of shallow mine workings and mine entries, spontaneous combustion of coal and unsealed abandoned mine entries permitting access into old mines. See website for further details: http://coal.decc.gov.uk/en/coal/cms/services/safety/24_hour/24_hour.aspx

Countryside and Rights of Way Service,
City of Bradford Metropoliton District Council

www.bradford.gov.uk/bmdc/the_environment/countryside_and_rights_of_way

telephone: 01274 432425

Countryside Services, Calderdale Council

www.calderdale.gov.uk/environment/countryside/conservation/index.html

email: countryside@calderdale.gov.uk

telephone: 0845 245 6000

Maps

We advise that you carry the correct map when exploring this landscape, all the trails can be followed using this publication and Ordnance Survey Explorer maps OL21 (Oxenhope and Todmorden) and 288 (Baildon).

High Plain Delf on the summit of Baildon Hill

Baildon Moor heritage trail

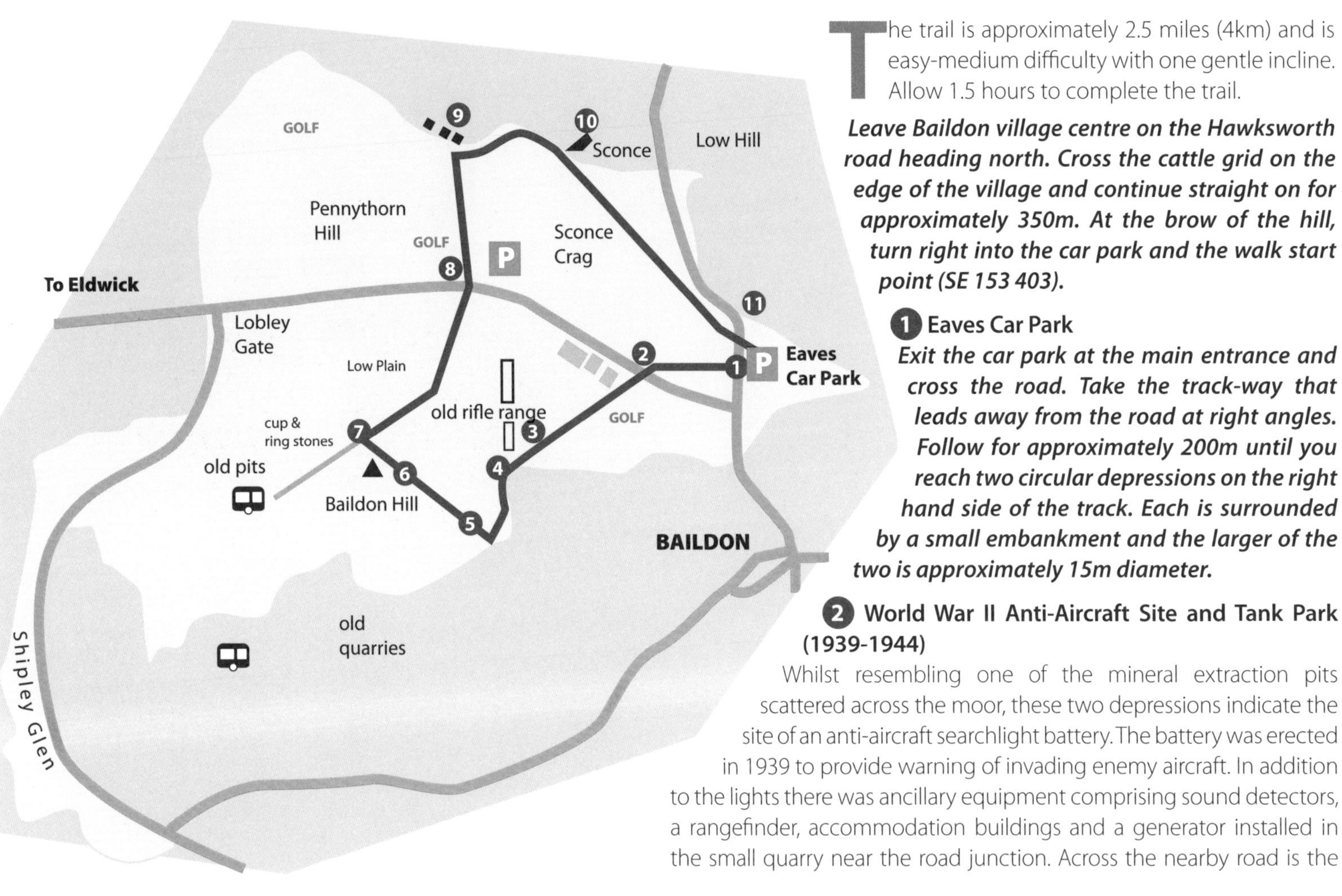

The trail is approximately 2.5 miles (4km) and is easy-medium difficulty with one gentle incline. Allow 1.5 hours to complete the trail.

Leave Baildon village centre on the Hawksworth road heading north. Cross the cattle grid on the edge of the village and continue straight on for approximately 350m. At the brow of the hill, turn right into the car park and the walk start point (SE 153 403).

1 Eaves Car Park

Exit the car park at the main entrance and cross the road. Take the track-way that leads away from the road at right angles. Follow for approximately 200m until you reach two circular depressions on the right hand side of the track. Each is surrounded by a small embankment and the larger of the two is approximately 15m diameter.

2 World War II Anti-Aircraft Site and Tank Park (1939-1944)

Whilst resembling one of the mineral extraction pits scattered across the moor, these two depressions indicate the site of an anti-aircraft searchlight battery. The battery was erected in 1939 to provide warning of invading enemy aircraft. In addition to the lights there was ancillary equipment comprising sound detectors, a rangefinder, accommodation buildings and a generator installed in the small quarry near the road junction. Across the nearby road is the present golf course. This was used as a tank training ground which extended over the hill to Glen Road. Tanks were parked and stored in the area between the Eaves Car Park and the searchlight battery.

Continue along the track and cross over the road. Walk across the open moor heading towards the far bottom corner of the reservoir's boundary wall. From here continue across the moor and the golf course, avoiding the greens and tees, in the direction of the houses. Head towards the right corner of the dry-stone boundary wall at the base of the hillside.

3 Victorian Rifle Range

From the wall corner and looking back along the base of the hillside towards open country you can see what appears to be terracing and trenching along the base of the slope. This is the remains of a line of target butts at the end of a military rifle range which was used by Victorian Rifle Volunteers and early twentieth century Territorial Army recruits, many of whom were the first to volunteer for service in 1914. Shooting took place over a range of distances up to 800yds (approximately 750m), shooting from near the Baildon-Hawksworth Road. From the range's beginnings in 1859 shooting competitions were common events. In 1861 Sir Titus Salt presented a first prize of a rifle and case valued at 25 guineas. The current seventeenth and eighteenth fairways were constructed in 1925 when the golf course was redesigned following the earlier decommissioning of the ranges.

Remain near the wall corner and look away from Baildon towards the open country a few metres in front of you - and behind the seventeenth green/eighteenth tee - is a cutting into the hillside.

4 Fireclay and coal mine

This is the entrance to a late nineteenth/early twentieth century coal/fireclay mine. This entrance, or 'adit', was dug into the hillside and followed the higher of two coal seams which run through Baildon Hill, known as the 'Hard Bed'. As the strata are almost horizontal this seam can be found outcropping along the contour of the hillside. Fireclays are sedimentary mudstones that occur as seat-earths which underlie almost all coal seams and represent the fossil soils on which the coal forming vegetation once grew. Their mineral constituents of mica and fine-grained quartz mean they are able to resist heat, which has led to their use in the manufacture of furnace linings.

This mine was probably worked mainly for fireclay rather than coal as it was operated by the Yorkshire Ganister Company which had a factory at Baildon Green producing fire-bricks. The mine was closed in the early years of the twentieth century and the Company ceased operating at Baildon Green a few years later.

Keeping the wall to your left hand side, follow it up the hill until you come to a bench seat.

5 Cinder Rocks and Brantcliffe Pit

The large mounds of cinder are the burnt remains of colliery waste and poor quality coal which was burnt on site to produce material for road construction. It is generally understood that there were three main coal pits on Baildon Hill operating in the mid-nineteenth century. One at Lobley Gate on the road to Eldwick near the cattle grid, one on Low Plain near Dobrudden Farm and Brantcliffe Pit, where you are now. These moderately sized pits operated at depth exploiting the 'Soft Bed' coal seam which is the lower of the two seams running through Baildon Hill.

Follow the path up the hill keeping the wall to your left hand side. At the top of the short incline the ground starts to flatten out. Here you leave the wall, and Baildon, behind you and head to your right towards the higher ground. On reaching the flat top of the hill, 'High Plain', head towards the white concrete pillar and nearby stone pillar, skirting around the various pits and depressions, many of which are the water-filled remains of clay workings.

6 Triangulation Point

This white-painted concrete pillar marks the top of Baildon Hill. 'Trig points', as they are more commonly known, were first erected by the Ordnance Survey in 1935 from a design by Brigadier Martin Hotline, then head of the Trigonometrical and Levelling Division of the Ordnance Survey. Each pillar is a fixed base of precisely known location and altitude onto which a theodolite can be fixed. They were originally used for the re-triangulation survey of Great Britain. Whilst some are still used today for their original purpose, many have been removed and others serve only as a guide to walkers. Modern satellite technology has superseded their original purpose. Sited close by is a directional diorama sponsored and erected by 'The Friends of Baildon Moor'.

Standing at the diorama pillar leave 'High Plain' heading in a north-westerly direction. As you start to go downhill you will notice a small gulley-like feature leading down to the stoned track adjacent to which is a large cinder heap. Continue downhill until you reach this stoned track.

7 Cinder Heap and Chimney Flue

This extensive flat area is known as 'Low Plain' and was the site of one of the three major nineteenth century coal pits. Around the cinder

Illustration of the Cinder Heap

heap were many buildings associated with the pit, and the gulley feature is the flue from the boilers which led to a chimney on the hilltop. Whilst coal waste was burnt to produce road construction material it has been suggested that these cinder heaps are a result of an attempt to turn un-saleable coal into a saleable construction material which also failed as a result of a declining market. Local reports talk of the site burning for days and being visible from some considerable distance.

Leave the cinder heap with Dobrudden Farm caravan site behind you. Walk along the stoned track until you reach the Baildon–Eldwick Road. Cross over the road into the car park.

8 Bronze Age Burial Mound

At the right hand side of the car park, when facing the road, you can see earthworks in the form of two parallel lines. These embankments form the corner of what appears to be a square or parallelogram. In 1880 these embankments were recorded as being almost 120ft (approximately 40m) in length, 20ft (approximately 5m) in width and almost 3.5ft (approximately 1m) in height. Whilst there is no evidence of the structure's original form, numerous 3500-4500 year old Bronze Age cinerary urns of varying styles were recovered from adjacent sites in the latter half of the nineteenth century. This is consistent with the period of many other prehistoric artefacts recovered from the moors area. The urns are displayed in local museums.

From the back of the car park, opposite the road, head in a northerly direction across the golf course. Walk downhill towards three small wooden buildings sited in the fields immediately behind a dry-stone boundary wall. Walk over to the buildings but do not cross the wall.

9 Three Wooden Buildings

These wooden buildings are the remains of a group of 14 such properties which were erected in the post-World War I period and owned by separate individuals for use as weekend or holiday cottages. Throughout the late nineteenth and early twentieth century this whole moorland area was a popular weekend and holiday retreat for people escaping built-up city areas. At peak holiday times there are records of tens of thousands of people using the tramway and fairground rides and amusements that the Glen area had to offer. Located further downstream of the small beck was a separate and slightly larger building known as 'Aunt Aggies' café. This was a popular spot with walkers, selling teas and refreshments, until it was eventually demolished in about 1970

Keep the wall on your left hand side and follow it until you *reach a small metal gate leading from the moor to what is now the Scout campsite.*

An artist's impression of the hamlet of Sconce, Bingley parish, in the early twentieth century, which was demolished in 1934/35 and had mining associations

10 Sconce Scout Camp Site

This is the former site of the enclosed hamlet of Sconce which comprised 13 cottages and outbuildings. It was constructed in 1730–50, possibly on behalf of the Ferrands family as dwellings for miners working the coal seams on the moor. The hamlet had many owners and its tenants many trades over its lifetime before it was finally demolished in 1934/5 on the grounds of lack of sanitation and services. The site was purchased in 1964 by Shipley and Baildon Scout Council for use as a camping site and training centre. The metal gateway and entrance to the site is an original access way into the hamlet from a track, suitable for a horse and cart, which ran across the moor.

Leave Sconce site and follow the newly laid stone pathway across the moor noting the many depressions either side which are remains of the pits dug to gain access to the coal seam. Follow the pathway to its termination in a small car park alongside the Baildon – Hawkesworth road.

11 Sandstone Quarry

Overlooking the car park is a small rock face which is the end of Eaves Quarry. The quarry face extends in a straight line for over 100m towards Baildon and is one of many quarries sited at the edge of the moor. The geology of Baildon Hill comprises almost horizontal, alternating beds of sandstone and mudstone, within which lie coal seams, ganister and fireclay deposits. Unlike these deposits the sandstone was extracted from open quarries and was used for building and road construction material. Careful study and observation of the exposed rock faces which remain can tell us much of the formation of the rocks. They were formed 320 million years ago in what is referred to as the Upper Carboniferous period. Many of the rock forming sediments were laid in delta areas and the coals seams were formed as a result of a long process of decay and compression of the complex forest systems that grew on them.

Cross the road, taking care of the close bend, and follow the track along the edge of the hilltop back to the Eaves car park.

Different levels in the quarry produced material of differing qualities each suited to different construction methods. The Eaves Quarry produced all the walling stone, roofing slates, floor flags and even the stone sinks used to build the nearby hamlet of Moorside, and other more prestigious buildings in Baildon such as the Parish Church of St John. Moorside comprised a cluster of 23 properties sited alongside the track at the base of the quarry. The 1841 census records Moorside as having a population of 131 people. It is believed to have dated back to the mid-eighteenth century but was demolished in the mid-1960s on the grounds of being unfit for human habitation.

Quarry at Eaves Crag, Baildon Moor

Reconstruction illustration of Moorside

Oxenhope Moor heritage trail

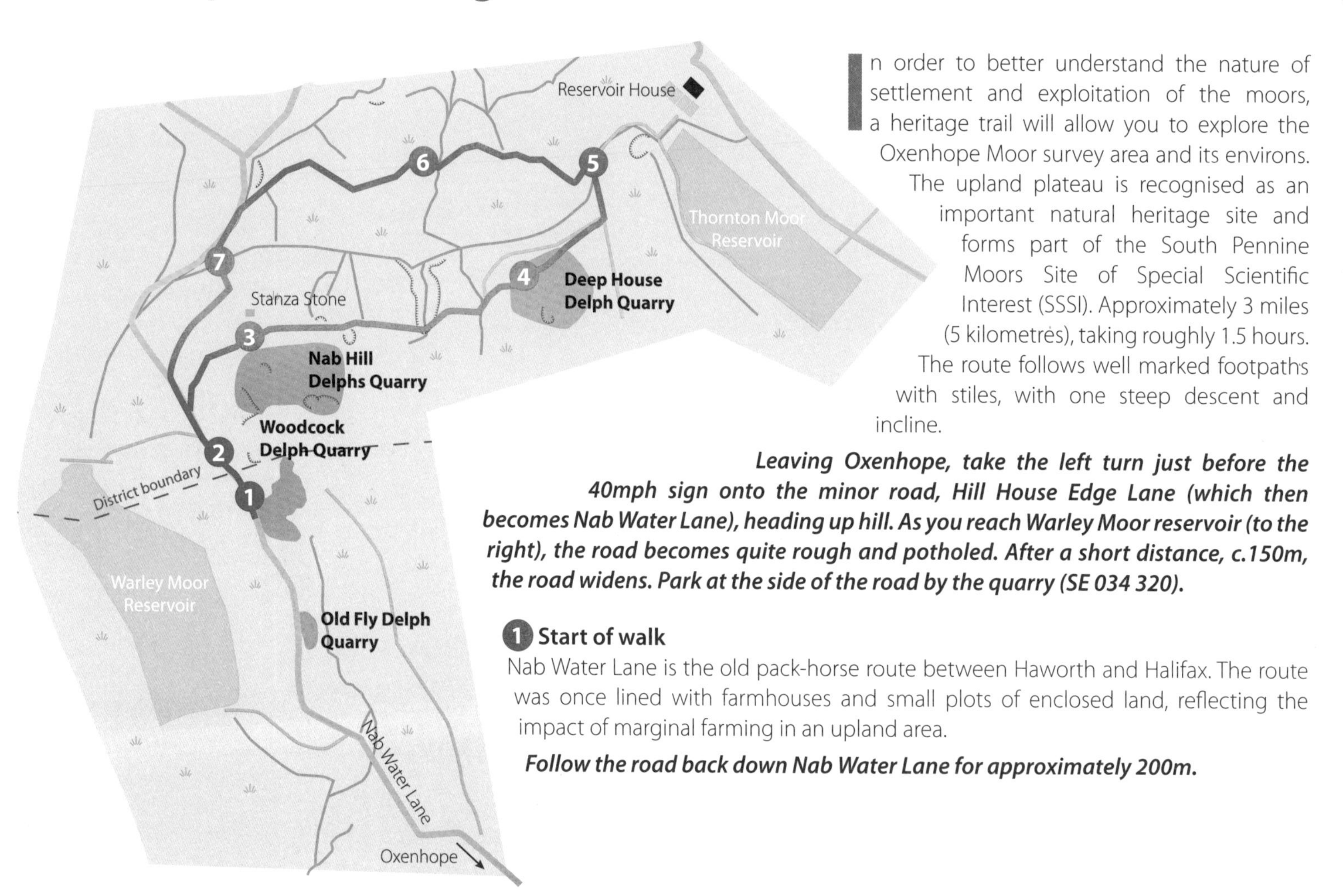

In order to better understand the nature of settlement and exploitation of the moors, a heritage trail will allow you to explore the Oxenhope Moor survey area and its environs. The upland plateau is recognised as an important natural heritage site and forms part of the South Pennine Moors Site of Special Scientific Interest (SSSI). Approximately 3 miles (5 kilometres), taking roughly 1.5 hours. The route follows well marked footpaths with stiles, with one steep descent and incline.

Leaving Oxenhope, take the left turn just before the 40mph sign onto the minor road, Hill House Edge Lane (which then becomes Nab Water Lane), heading up hill. As you reach Warley Moor reservoir (to the right), the road becomes quite rough and potholed. After a short distance, c.150m, the road widens. Park at the side of the road by the quarry (SE 034 320).

1 Start of walk

Nab Water Lane is the old pack-horse route between Haworth and Halifax. The route was once lined with farmhouses and small plots of enclosed land, reflecting the impact of marginal farming in an upland area.

Follow the road back down Nab Water Lane for approximately 200m.

2 Fly hamlet

A line of ruined farmsteads is just visible to the left-hand side of the road (the example on page 17 can be seen from SE 033 322). These once formed the isolated upland settlement at Fly. To the west of the road at Old Fly, a temporary camp was built for the construction of the Fly Flats or Warley Moor reservoir. Approximately 150m along the road in the passing place to the right, erosion of the road metalling has begun to reveal several large stone slabs quarried from the local delphs.

The walk passes the extensive tips of the stone delphs on the western edge of Nab Hill escarpment. The oldest of these have now been lost to modern working but some of their infrastructure remains in the immediate area.

Continue down the road until you reach a kissing gate to the right. Follow the footpath through the kissing gate, heading up hill. The path takes you through the now grassed over quarry waste which is retained by walls made form waste stone – termed 'Judd Walls'.

3 Nab Hill delphs

For several centuries the domestic production of textiles dominated the upper valley sides in the South Pennines. But the nature of the land also allowed other industries to develop. The immediate area is rich in mineral resources: building stone in the north and west of the area; and seams of coal and clay to the east where the Denholme Clough Fault marks the north-west extremity of the Yorkshire Coalfield. All these resources could be exploited commercially by small individual concerns often employing younger male members of local families as the domestic textile industry declined. Their extraction enabled in some way the industrialisation of the region.

The quarry here, at the summit, was the focus of the work carried out by the Oxenhope Moor survey team.

Along the escarpment edge to the north, just passed the quarry and before the cairn, is the 'Mist' Stanza Stone (see it on page 57 and at SE 033 327), one of a series of stones across the South Pennine landscape inscribed with a stanza written by Simon Armitage (for more information see http://www.ilkleyliteraturefestival.org.uk/test/stanza-stones-trail-guide/).

After exploring the quarry and the Stanza Stone, continue along the path, heading east.

4 Deep House delphs

The area of Deep House Delphs allows extensive views to the Yorkshire Dales to the north.

Continue along the path until it joins a path aligned north-south, before you reach Thornton Moor reservoir visible in front of you. Turn left and continue for a short distance (c.200m) along the path.

5 Thornton Moor reservoir and reservoir house

The present Reservoir House at Thornton Moor, near to the filter beds at the west side of the reservoir, is built on the site of several 'lodging houses', temporary accommodation built to house some of the large labour force required for the construction of the waterworks. Looking north, at the east end of the hill lay the settlement of Sawood once the location of Oxenhope's largest colliery, and beyond Bradshaw Head the site of former clay pits and a brickworks.

Turn left (heading west) along the Calder Aire Link bridleway. Initially, the bridleway follows Sawood Lane.

❻ Workers' housing

The remains of several 'lost' hamlets below the escarpment, home to farmers, textile workers, quarry workers and coal miners in former times are also visible. Freeholders had rights to take the peat for fuel and small amounts of building stones for their own use or to repair communal roads. Farming at this altitude on small plots of allotted enclosures was difficult due to the poor marginal nature of the land and the climate. However, the edges of the moorland were successfully settled due to the adoption of the 'dual economy'; the supplementing of farming incomes with low levels of industrial activity.

Continue following the Calder Aire Link bridleway.

❼ Reservoir conduits

The route takes you close to and occasionally crosses conduits – channels for collecting and conveying water to reservoirs. The landscape was dramatically altered as local Municipal Boroughs found the moorland edges and tops ideal for water catchment schemes to provide water for domestic and industrial demand in the burgeoning industrial towns of the valley bottoms.

When the bridleway joins Nab Water Lane, turn left for a short distance. Take the footpath to the left that crosses the conduit via a footbridge, then follow the footpath to the right, skirting along the side of the hill, rejoining Nab Water Lane. From here, continue up Nab Water Lane, once again passing the ruined farmsteads and Warley Moor reservoir.

Before you leave, pause for a moment to take in the view across Midgley and Warley Moors, towards Stoodley Pike in the distance. The extent of our interaction with the upland area between Oxenhope and Warley can still be recognised on the ground despite the landscape appearing empty and barren. Imagine the area during the height of industrialisation during the nineteenth century – a contrast to the area today.

A section of the Thornton Moor Reservoir conduit

The 'Mist' Stanza Stone inscribed with a poem by Simon Armitage

Todmorden Moor geology and heritage trail

A Todmorden Moor Geology and Heritage Trail can be accessed from the Todmorden to Bacup road (A681). Take the turn marked Sourhall for 3/4 mile, turn left over the cattle grid and just under 300m turn left onto a rough track, which is Flowerscar Road. Park at the side of the track. The trail is signposted from grid reference SD 909 251.

Interpretation board at the start of the Trail

Spoil tip at Sandy Road Colliery, Todmorden Moor

Further Information

Societies and organisations

Association for Industrial Archaeology
http://industrial-archaeology.org/

Coal Authority
http://coal.decc.gov.uk/

Friends of Baildon Moor
http://baildonmoor.org/wordpress/

National Association of Mining History Organisations
http://www.namho.org.uk/

Northern Mine Research Society
http://www.nmrs.org.uk/

Todmorden Moor Restoration Trust
http://todmordenmoor.org.uk/

West Yorkshire Geology Trust
http://www.wyorksgeologytrust.org/

Yorkshire Archaeological Society, Industrial History Section
http://www.industrialhistory.yas.org.uk/

Museums

Bradford Industrial Museum
Moorside Mills, Moorside Road, Eccleshill, Bradford, BD2 3HP

http://www.bradfordmuseums.org/venues/industrialmuseum/index.php

National Coal Mining Museum
Caphouse Colliery, New Road, Overton, Wakefield WF4 4RH

http://www.ncm.org.uk/

Publications

Coyle, G. 2010. *The Riches Beneath our Feet*. Oxford: Oxford University Press

Fieldhouse, J. 1978. *Bradford*. 3rd edition. Bradford: City of Bradford Metropolitan Council

Commons and Lords Hansard. *1842. Mines and Collieries Bill* 1842 vol 65. From

http://hansard.millbanksystems.com/lords/1842/jul/14/mines-and-collieries#s3v0065p0_18420714_hol_51

Palmer, M., Nevell, M. and Sissons, M. 2012. *Industrial Archaeology. A handbook*. CBA Practical Handbook 21. York: Council for British Archaeology.

Rothwell, M. 2010. *Industrial Heritage - A Guide to the Industrial Archaeology of Bacup and Stacksteads*. Bridgestone Press